# BEYOND GEOMETRY

# THE HISTORY OF MATHEMATICS

# BEYOND GEOMETRY

## A NEW MATHEMATICS OF SPACE AND FORM

*John Tabak, Ph.D.*

Facts On File
*An Infobase Learning Company*

**BEYOND GEOMETRY: A New Mathematics of Space and Form**

Copyright © 2011 by John Tabak, Ph.D.

Facts On File, Inc.
An imprint of Infobase Learning
132 West 31st Street
New York NY 10001

**Library of Congress Cataloging-in-Publication Data**

Tabak, John.
  Beyond geometry: a new mathematics of space and form / John Tabak.
    p. cm.—(The history of mathematics)
  Includes bibliographical references and index.
  ISBN 978-0-8160-7945-2
  1. Topology. 2. Set theory. 3. Geometry. I. Title.
  QA611.T33 2011
  516—dc22        2010023887

Facts On File books are available at special discounts when purchased in bulk quantities for businesses, associations, institutions, or sales promotions. Please call our Special Sales Department in New York at (212) 967-8800 or (800) 322-8755.

You can find Facts On File on the World Wide Web at http://www.infobaselearning.com

Excerpts included herewith have been reprinted by permission of the copyright holders; the author has made every effort to contact copyright holders. The publishers will be glad to rectify, in future editions, any errors or omissions brought to their notice.

Text design by David Strelecky
Composition by Hermitage Publishing Services
Illustrations by Dale Williams
Photo research by Elizabeth H. Oakes
Cover printed by Yurchak Printing, Inc., Landisville, Pa.
Book printed and bound by Yurchak Printing, Inc., Landisville, Pa.
Date printed: May 2011
Printed in the United States of America

10 9 8 7 6 5 4 3 2 1

This book is printed on acid-free paper.

*For Stuart Dugowson.*
*He was a generous spirit with an inquiring mind.*

# CONTENTS

# PREFACE

Of all human activities, mathematics is one of the oldest. Mathematics can be found on the cuneiform tablets of the Mesopotamians, on the papyri of the Egyptians, and in texts from ancient China, the Indian subcontinent, and the indigenous cultures of Central America. Sophisticated mathematical research was carried out in the Middle East for several centuries after the birth of Muhammad, and advanced mathematics has been a hallmark of European culture since the Renaissance. Today, mathematical research is carried out across the world, and it is a remarkable fact that there is no end in sight. The more we learn of mathematics, the faster the pace of discovery.

Contemporary mathematics is often extremely abstract, and the important questions with which mathematicians concern themselves can sometimes be difficult to describe to the interested nonspecialist. Perhaps this is one reason that so many histories of mathematics give so little attention to the last 100 years of discovery—this, despite the fact that the last 100 years have probably been the most productive period in the history of mathematics. One unique feature of this six-volume History of Mathematics is that it covers a significant portion of recent mathematical history as well as the origins. And with the help of in-depth interviews with prominent mathematicians—one for each volume—it is hoped that the reader will develop an appreciation for current work in mathematics as well as an interest in the future of this remarkable subject.

*Numbers* details the evolution of the concept of number from the simplest counting schemes to the discovery of uncomputable numbers in the latter half of the 20th century. Divided into three parts, this volume first treats numbers from the point of view of computation. The second part details the evolution of the concept of number, a process that took thousands of years and culminated in what every student recognizes as "the real number line," an extremely important and subtle mathematical idea. The third part of this volume concerns the evolution of the concept of the infinite. In particular, it covers Georg Cantor's discovery (or creation, depending on one's point of view) of transfinite numbers and his efforts to place set theory at the heart of modern mathematics. The most important ramifications of Cantor's work, the attempt to axiomatize mathematics carried out by David Hilbert and Bertrand Russell, and the discovery by Kurt Gödel and Alan Turing that there are limitations on what can be learned from the axiomatic method, are also described. The last chapter ends with the discovery of uncomputable numbers, a remarkable consequence of the work of Kurt Gödel and Alan Turing. The book concludes with an interview with Professor Karlis Podnieks, a mathematician of remarkable insights and a broad array of interests.

*Probability and Statistics* describes subjects that have become central to modern thought. Statistics now lies at the heart of the way that most information is communicated and interpreted. Much of our understanding of economics, science, marketing, and a host of other subjects is expressed in the language of statistics. And for many of us statistical language has become part of everyday discourse. Similarly, probability theory is used to predict everything from the weather to the success of space missions to the value of mortgage-backed securities.

The first half of the volume treats probability beginning with the earliest ideas about chance and the foundational work of Blaise Pascal and Pierre Fermat. In addition to the development of the mathematics of probability, considerable attention is given to the application of probability theory to the study of smallpox and the misapplication of probability to modern finance. More than most branches of mathematics, probability is an applied discipline, and its uses and misuses are important to us all. Statistics is the subject of the second half of the book. Beginning with the earliest examples of statistical thought, which are found in the writings of John Graunt and Edmund Halley, the volume gives special attention to two pioneers of statistical thinking, Karl Pearson and R. A. Fisher, and it describes some especially important uses and misuses of statistics, including the use of statistics in the field of public health, an application of vital interest. The book concludes with an interview with Dr. Michael Stamatelatos, director of the Safety and Assurance Requirements Division in the Office of Safety and Mission Assurance at NASA, on the ways that probability theory, specifically the methodology of probabilistic risk assessment, is used to assess risk and improve reliability.

*Geometry* discusses one of the oldest of all branches of mathematics. Special attention is given to Greek geometry, which set the standard both for mathematical creativity and rigor for many centuries. So important was Euclidean geometry that it was not until the 19th century that mathematicians became willing to consider the existence of alternative and equally valid geometrical systems. This 19th-century revolution in mathematical, philosophical, and scientific thought is described in some detail, as are some alternatives to Euclidean geometry, including projective geometry, the non-Euclidean geometry of Nikolay Ivanovich Lobachevsky and János Bolyai, the higher (but finite) dimensional geometry of Riemann, infinite-dimensional geometric ideas, and some of the geometrical implications of the theory of relativity. The volume concludes with an interview with Professor Krystyna Kuperberg of Auburn University about her work in geometry and dynamical systems, a branch of mathematics heavily dependent on ideas from geometry. A successful and highly insightful mathematician, she also discusses the role of intuition in her research.

Mathematics is also the language of science, and mathematical methods are an important tool of discovery for scientists in many disciplines. *Mathematics and the Laws of Nature* provides an overview of the ways that mathematical thinking has influenced the evolution of science—especially the use of deductive reasoning in the development of physics, chemistry, and population genetics. It also discusses the limits of deductive reasoning in the development of science.

In antiquity, the study of geometry was often perceived as identical to the study of nature, but the axioms of Euclidean geometry were gradually supplemented by the axioms of classical physics: conservation of mass, conservation of momentum, and conservation of energy. The significance of geometry as an organizing principle in nature was briefly subordinated by the discovery of relativity theory but restored in the 20th century by Emmy Noether's work on the relationships between conservation laws and symmetries. The book emphasizes the evolution of classical physics because classical insights remain the most important insights in many branches of science and engineering. The text also includes information on the relationship between the laws of classical physics and more recent discoveries that conflict with the classical model of nature. The main body of the text concludes with a section on the ways that probabilistic thought has sometimes supplanted older ideas about determinism. An interview with Dr. Renate Hagedorn about her work at the European Centre for Medium-Range Weather Forecasts (ECMWF), a leading center for research into meteorology and a place where many of the concepts described in this book are regularly put to the test, follows.

Of all mathematical disciplines, algebra has changed the most. While earlier generations of geometers would recognize—if not immediately understand—much of modern geometry as an extension of the subject that they had studied, it is doubtful that earlier generations of algebraists would recognize most of modern algebra as in any way related to the subject to which they devoted their time. *Algebra* details the regular revolutions in thought that have occurred in one of the most useful and vital areas of contemporary mathematics: Ancient proto-algebras, the concepts of algebra that originated in the Indian subcontinent and in the Middle East, the "reduction" of geometry to algebra begun by René Descartes, the abstract algebras that grew out of the work of Évariste Galois, the work of George Boole and some of the applications of his algebra, the theory of matrices, and the work of Emmy Noether are all described. Illustrative examples are also included. The book concludes with an interview with Dr. Bonita Saunders of the National Institute of Standards and Technology about her work on the Digital Library of Mathematical Functions, a project that mixes mathematics and science, computers and aesthetics.

New to the History of Mathematics set is *Beyond Geometry*, a volume that is devoted to set-theoretic topology. Modern mathematics is often divided into three broad disciplines: analysis, algebra, and topology. Of these three, topology is the least known to the general public. So removed from daily experience is topology that even its subject matter is difficult to describe in a few sentences, but over the course of its roughly 100-year history, topology has become central to much of analysis as well as an important area of inquiry in its own right.

The term *topology* is applied to two very different disciplines: set-theoretic topology (also known as general topology and point-set topology), and the very different discipline of algebraic topology. For two reasons, this volume deals almost exclusively with the former. First, set-theoretic topology evolved along lines that were, in a sense, classical, and so its goals and techniques, when viewed from a certain perspective, more closely resemble those of subjects that most readers have already studied or will soon encounter. Second, some of the results of set-theoretic topology are incorporated into elementary calculus courses. Neither of these statements is true for algebraic topology, which, while a very important branch of mathematics, is based on ideas and techniques that few will encounter until the senior year of an undergraduate education in mathematics.

The first few chapters of *Beyond Geometry* provide background information needed to put the basic ideas and goals of set-theoretic topology into context. They enable the reader to better appreciate the work of the pioneers in this field. The discoveries of Bolzano, Cantor, Dedekind, and Peano are described in some detail because they provided both the motivation and foundation for much early topological research. Special attention is also given to the foundational work of Felix Hausdorff.

Set-theoretic topology has also been associated with nationalism and unusual educational philosophies. The emergence of Warsaw, Poland, as a center for topological research prior to World War II was motivated, in part, by feelings of nationalism among Polish mathematicians, and the topologist R. L. Moore at the University of Texas produced many important topologists while employing a radical approach to education that remains controversial to this day. Japan was also a prominent center of topological research, and so it remains. The main body of the text concludes with some applications of topology, especially dimension theory, and topology as the foundation for the field of analysis. This volume contains an interview with Professor Scott Williams, an insightful thinker and pioneering topologist, on the nature of topological research and topology's place within mathematics.

The five revised editions contain a more comprehensive chronology, valid for all six volumes, an updated section of further resources, and many new color photos and line drawings. The visuals are an important part of each volume, as they enhance the narrative and illustrate a number of important (and very visual) ideas. The History of Mathematics should prove useful as a resource. It is also my hope that it will prove to be an enjoyable story to read—a tale of the evolution of some of humanity's most profound and most useful ideas.

# ACKNOWLEDGMENTS

The author thanks Professor Scott Williams for the generous way that he shared his time and insights and Ms. Atsuko Waki for her invaluable help with the research on which this book is based. Special thanks to Mr. Yasuhiro Morita for providing extensive biographical information and a number of highly insightful remarks on the life and work of his father, the late Kiiti Morita. Also important to the production of this volume were the suggestions of executive editor Frank K. Darmstadt and the careful photo research of Elizabeth Oakes.

# INTRODUCTION

For millennia, Euclidean geometry, the geometry of the ancient Greeks, set the standard for rigor in mathematics—it was the only branch of mathematics that had been developed axiomatically. Euclidean geometry, most philosophers and mathematicians agreed, was the language of mathematics. However, early in the 19th century, mathematicians developed geometries very different from Euclid's simply by choosing *axioms* different from those used by Euclid. These new geometries were internally consistent in the sense that mathematicians could find no theorems arising within these geometries that could be proved both true and false. Some of these alternative geometries even provided insight into certain aspects of physical space. Mathematicians and philosophers, who had longed believed that Euclid's geometry embodied mathematical truth, came to see Euclidean geometry as one possible axiomatic system among many. Nevertheless, the discipline of geometry, suitably broadened, retained its place at the center of mathematical thought until the last decades of the 19th century.

Nevertheless, geometry proved to be ill-suited to the demands placed upon it. *Analysis*, for example, the branch of mathematics that grew out of calculus, had originally been expressed in the language of geometry, but the attempt to express analytical ideas in geometric language led to logical difficulties. Conceptually, geometry was not rich enough to serve as a basis for the "new" mathematics that was developed during the latter half of the 19th century. Consequently, mathematicians began to seek another mathematical language in which to express their insights. They found it in the language of sets.

The theory of sets began with the study of sets of geometric points—points in space, points in the plane, and points on the line—but mathematicians soon discovered that their set-theoretic ideas applied just as well to sets of functions, sets of lines, and sets of planes. In fact, many mathematicians ceased to ascribe any extramathematical meaning at all to the term *point*. To them, points were just elements of sets, and they began to study "abstract point sets" endowed with a structure called a *topology*, another idea abstracted from the study of sets of geometric points. These sets of abstract points, each equipped with one or more topologies, came to be called *topological spaces*.

Many mathematicians turned their attention to the study of topological spaces and made many surprising and useful discoveries. The value of the research stemmed from the fact that set-theoretic language is so widely applicable. Discoveries about the properties of abstract sets also applied in more practical settings. Set-theoretic language became the language of mathematics, and along with algebra and analysis, topology became one of the three major branches of mathematics.

*Beyond Geometry* describes how *set-theoretic topology* developed and why it now occupies a central place in mathematics. Because those mathematical pioneers who developed set-theoretic topology were so classical in their outlook, there are close conceptual parallels between the development of set-theoretic topology and geometry. The first chapter of this book describes the axiomatic method as well as provides a definition of what a geometric property is—that is, what it is that mathematicians study when they study geometry. Finally, chapter 1 describes how early analysts incorporated geometric thinking into their development of the calculus.

Chapter 2 describes how classical geometric thinking could not adequately account for the many phenomena that 19th-century research revealed. Mathematicians, led by Georg Cantor and Richard Dedekind, responded by creating a theory of sets. Their struggle to develop a new conceptual framework for mathematics is described in chapter 3.

Cantor's and Dedekind's work was formalized by the early topologists, especially Felix Hausdorff. His abstract topological space is described in chapter 4, as is the concept of a topological property. Just as geometric properties can be defined as those properties that remain unchanged under geometric *transformations*, topological properties can be defined as exactly those properties that remain unchanged under topological transformations. *Euclidean transformations* are relatively easy to characterize; topological transformations are less intuitive but far more useful. While every Euclidean transformation is a topological transformation, most topological transformations are not Euclidean transformations, and so the properties that topological transformations preserve are quite different from Euclidean properties. Making these ideas precise is the goal of chapters 4 and 5.

Topology, more than most branches of mathematics, has also been associated with controversy. This has found its expression in highly individualistic schools of topology. The Polish school of topology was established shortly after the end of World War I when Poland was reconstituted as a nation. Polish mathematicians, eager to make their presence felt, focused their efforts on a few rapidly evolving branches of mathematics. Topology was one these. The school of topology founded by R. L. Moore at the University of Texas was famous because many topologists received their education there, and it was also famous for the method by which they were educated. Moore practiced what is best described as the Socratic method of education, and despite criticism of his ideas, his success as a teacher of higher mathematics has rarely been surpassed. The Japanese school was founded shortly after the end of World War II by Kiiti Morita, a highly successful researcher in the fields of algebra and topology. Self-taught in topology, he inspired a generation of Japanese mathematicians to focus their efforts on topological research.

Chapter 7 describes one of the subdisciplines of set-theoretic topology called *dimension theory*. Often described as one of the great accomplishments of set-theoretic topology, dimension theory makes explicit what it means to say that a space has one, two, three, or more generally, *n* dimensions. The problem of dimension is fundamental, and, surprisingly, the concept of dimension can be defined in several nonequivalent ways.

Chapter 8 describes why topology has become such an essential part of modern mathematical thought. The text concludes with an interview with Professor Scott Williams, a widely published and insightful topologist and independent thinker.

Topology is paradoxical. Perhaps more than any other major branch of mathematics, set-theoretic topology is inward looking in the sense that the questions that it asks and answers frequently have no connection with the physical world. Unlike calculus, in which questions are often phrased in terms of projectiles and the volumes of geometric objects, topology is concerned almost exclusively with highly abstract questions. Even so, topology has proved its value in many areas of mathematics, even calculus, and today, much of mathematics cannot be understood without topological ideas. It is hoped that this book will increase awareness of this vital field as well as prove to be a valuable resource.

# 1

# TOPOLOGY: A PREHISTORY

Topology is one of the newest of all the major branches of mathematics. The formal development of the subject began in earnest during the early years of the 20th century, and within a few decades, numerous and important discoveries advanced topology from its earliest foundational ideas to a high level of sophistication. It now occupies a central place in mathematical thought. Despite its importance, topology remains largely unknown (or at least unappreciated) outside mathematics. By contrast, algebra and geometry are encountered by practically everyone. When reading articles about science and math, for example, most readers are not surprised to encounter equations involving variables (algebra), diagrams (geometry), and perhaps even *derivatives* and integrals (calculus). It may not always be entirely clear what a particular symbol or diagram means, but our experience has taught us to expect them, and our education has enabled us to extract some meaning from them. It is still a common mistake, however, to confuse topology with topography.

Despite its relative obscurity, set-theoretic topology could not be more important to contemporary mathematics. Topological results provide much of the theoretical underpinnings for calculus, for example, as well as for all the mathematics that arose from the study of calculus. (The branch of mathematics that arose from calculus is called analysis.) There are many other applications for topology as well, some of which are described later in this book. To appreciate what topologists study, how they study it, and why their work is important, it helps to begin by looking at the very early history of mathematics. This chapter covers three topics that are of special significance in understanding topology: the

*Temple of Apollo. The ancient Greeks were the first to investigate mathematics axiomatically. They set a standard for rigorous thinking that was not surpassed until the 19th century.* (Ballista)

axiomatic method, the idea of a transformation, and the concept of continuity.

## Euclid's Axioms

The mathematics that was developed in ancient Greece occupies a unique place in the history of mathematics, and for good reason. From a modern point of view, Greek mathematics is special because of the way they developed mathematical knowledge. In fact, from a historical point of view, the way the Greeks investigated mathematics is probably more important than any discoveries that they made about geometry, and the Greeks understood the importance of their method. Referring to Thales of Miletus (ca. 625–ca. 547 B.C.E.), often described as the first Greek mathematician, the philosopher Aristotle (384–322 B.C.E.) remarked, "To Thales, the primary question was not what do we know, but how do we know it."

Here is one reason that the method used to investigate mathematics is so important: Suppose that we are given two statements—call them statement $A$ and statement $B$—and suppose that we can prove that the truth of statement $A$ implies the truth of statement $B$. In other words, we can prove that $B$ is true provided $A$ is true. Such a proof does not address the truth of $B$. Instead, it shifts our attention from $B$ to $A$. If $A$ is true *then* $B$ is true, but is $A$ true? Rather than proving the truth of $B$, we have begun a backward chain of logical implications. The next step is to attempt to establish the truth of $A$ in terms of some other statement. The situation is similar to the kinds of conversations that very young children enjoy. First, the child will ask a question. The adult will answer, and the child responds with "Why?" Each succeeding answer elicits the same why-response. For children, there is no satisfactory answer. For mathematicians, the answer is a collection of statements called axioms, and the ancient Greeks were the first to use what is now known as the axiomatic method. The Greek dependence on the axiomatic method is best illustrated in the work of Euclid of Alexandria, the author of *Elements*, one of the most influential books in history.

The Greek mathematician Euclid (fl. 300 B.C.E.) lived in Alexandria, Egypt, and wrote a number of books about mathematics, some of which survive to the present day. Little is known of Euclid's personal life. Instead, he is remembered as the author of *Elements*, a textbook that reveals a great deal about the way the Greeks understood mathematics. Euclid's treatment of geometry in this textbook formed the basis for much of the mathematics that subsequently developed in the Middle East and Europe. It remained a central part of mathematics until the 19th century.

In *Elements*, Euclid begins his discussion of geometry with a list of undefined terms and definitions, and these are followed by a set of *axioms* and *postulates*, which together form the logical framework of his geometry. The undefined and defined terms constitute the vocabulary of his subject; the axioms and postulates describe the basic properties of the geometry. To put it another way: The axioms and postulates describe the basic relations that exist among the defined and undefined terms. The theorems, which constitute the main body of the work, are the logical deductions drawn from this conceptual framework. (Euclid lists many theorems in his work. Many other theorems were discovered later. Some theorems of Euclidean geometry remain to be discovered.) Because the theorems are logical consequences of the axioms and postulates, Euclidean geometry is entirely determined once the axioms and postulates are listed. In a logical sense, therefore, Euclid's geometry was completely determined when Euclid finished writing his axioms and postulates. The axioms and postulates are the final answer to why a statement is true in Euclidean geometry: A statement in Euclidean geometry is true because it is a logical consequence of Euclid's axioms and postulates.

It is now possible to describe what it is that mathematicians "discover" when they discover new mathematics: They deduce new and nonobvious conclusions from the axioms and postulates that define the subject. Strictly speaking, therefore, Euclid's successors do not create new mathematics at all. Instead, they reveal unexpected consequences of old axioms and postulates. Logically speaking, the consequences were already present for all to see. Mathematicians simply draw our attention to them.

The situation is otherwise in other branches of knowledge. Logical thinking also characterizes scientific and engineering research, of course, but in science and engineering, all results, no matter how logically derived, must also agree with experimental data. Outside of mathematics, experiment takes precedence over logic. Only in mathematics is truth entirely dependent on rules of logical inference. This is part of what makes mathematics different from other forms of human endeavor.

Euclid's axioms and postulates are as follows:

Axioms

1. *Things which are equal to the same thing are also equal to one another.*

2. *If equals be added to equals, the wholes are equal.*

3. *If equals be subtracted from equals, the remainders are equal.*

4. *Things which coincide with one another are equal to one another.*

5. *The whole is greater than the part.*

Postulates

1. *To draw a straight line from any point to any point*

2. *To produce a finite straight line continuously in a straight line*

3. *To describe a circle with any center and distance*

4. *That all right angles are equal to one another*

5. *That if a straight line falling on two straight lines makes the interior angles on the same side less than two right angles, the straight lines, if produced indefinitely, will meet on that side on which the angles are less than two right angles*

Euclid believed that his axioms—he also called them "common notions"—were somehow different from his postulates because he perceived axioms as more a matter of common sense than the conceptually more difficult postulates, but today all 10 statements are usually called axioms, with no distinction made between the terms *axioms* and *postulates* because, logically speaking, there is no distinction to make. (Postulates 1–3 mean that in Euclidean geometry, it is always possible to perform the indicated operations. It is, for example, always possible to draw a straight line connecting any two points [postulate 1], and given a straight line segment, it is always possible to lengthen or "produce" it still further [postulate 2], and given any point and any distance, it is always possible to draw a circle with center at that point and with the given distance as radius [postulate 3].)

Euclid could have chosen other axioms. Another mathematician might, for example, imagine a geometry that contained a pair of points that cannot be connected by a straight line. Logically, nothing prevents the creation of a geometry with this property, but this alternative geometry would not be the geometry of Euclid, because according to postulate 1, in Euclidean geometry *every* pair of points can be connected by a straight line.

While mathematicians now recognize that there is some freedom in the choice of the axioms one uses, not any set of statements can serve as a set of axioms. In particular, every set of axioms must be logically consistent, which is another way of saying that it should not be possible to prove a particular statement simultaneously true *and* false using the given set of axioms. Also, axioms should always be logically independent—that is, no axiom should be a logical consequence of the others. A statement that is a logical consequence of some of the axioms is a theorem, not an axiom. Euclid's fifth postulate provides a good illustration of these requirements.

All five of Euclid's axioms and the first four of his postulates are short and fairly simple. The fifth postulate is much wordier and grammatically more involved. In modern language it can be expressed as follows: Given any line and a point not on the line, there exists exactly one line through the given point that does not intersect the given line. (Put still another way: Given a line and a point not on the line, there is a unique second line passing through the given point that is parallel to the given line.) For 2,000 years after Euclid, many mathematicians—although not, evidently, Euclid—believed that the fifth postulate was redundant. They thought that it should be possible to prove the truth of the fifth postulate using Euclid's five axioms and the first four of his postulates. In more formal language, they thought that the fifth postulate was not logically independent of the other four postulates and five axioms. They were unsuccessful in their attempts to demonstrate that Euclid had made a mistake in including the fifth postulate—that is, they were unsuccessful in proving that the fifth postulate was a logical consequence of the other axioms and postulates. It was not until the 19th century that mathematicians were finally able to demonstrate that Euclid was correct. They showed that one cannot deduce the fifth postulate from Euclid's other axioms and postulates. Many historians believe that their demonstration marked the beginning of the modern era of mathematics (see chapter 2).

# Euclidean Transformations

The other aspect of Euclidean geometry that is especially note-worthy from a modern point of view is Euclid's use of transformations. While he did not give the matter as much attention as the axiomatic method, the idea of a transformation was important enough for him to include the idea (at least obliquely) in axiom 4:

Things which coincide with one another are equal to one another.

In Euclidean geometry, to test whether two geometric figures are equal—that is, to test whether they *coincide*—move one figure to the location occupied by the other and try to place it so that corresponding parts "match up." If all the corresponding parts can be made to match, the two figures are equal. Euclid was not quite as precise about this notion of equality as modern mathematicians are, and today this is recognized as something of an error on Euclid's part. Precision about the definition of motion is necessary because the types of motions that one allows determine the resulting geometry. Three types of motions are generally permitted within Euclidean geometry: translations, rotations, and reflections. A planar figure—a triangle, for example—is translated if it is moved within the plane so that it remains parallel to itself. A rotation means it is turned (rotated) about a point (any point) in the plane. One way to visualize a reflection is to imagine folding the plane along a line until the two halves of the plane coincide. The old triangle determines a new reflected triangle on the other half of the plane. The new triangle is the mirror image of the original. The distance between any two points of the new triangle is the same as the distance between the corresponding points of the original, but the orientation of the new triangle is opposite to that of the old one.

To see why it is important to specify exactly what types of motions one is willing to accept, imagine an isosceles triangle, which is a triangle that has two sides of equal length. Cut the triangle in half along its line of symmetry. Are the two halves congruent? If one allows reflections, the answer is yes. Reflecting across the isosceles triangle's line of symmetry demonstrates that each triangle is the

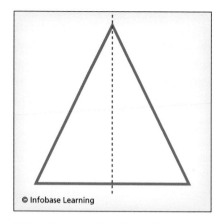

© Infobase Learning

*There is no combination of translations and rotations that will make the triangle on the left coincide with the triangle on the right. If reflections are allowed, they can be made to coincide by reflecting about the axis of symmetry of the isosceles triangle. The definition of congruence depends on the definition of an allowable transformation.*

mirror image of the other. (See the accompanying diagram.) However, if one does not allow reflections, they are not congruent—that is, they cannot be made to coincide. No collection of translations and rotations will enable one to "match up" the two halves. The set of allowable motions determines the meaning of the word *equal*. (Today we say *congruent* rather than *equal*.)

Translations, rotations, and reflections are examples of what mathematicians call transformations. These particular transformations preserve the measures of angles and the lengths of the figures on which they act. They are sometimes called rigid body motions. They are also sometimes called the set of Euclidean transformations, although Euclid did not know them by that name (nor did he call his geometry *Euclidean*).

One can characterize Euclidean geometry in one of two equivalent ways: First, Euclidean geometry is the set of logical deductions made from the axioms and postulates listed by Euclid. Alternatively, one can say Euclidean geometry consists of the study of those properties of figures (and only those properties of figures!) that remain unchanged under the set of Euclidean transformations. It is, again, important to keep in mind that if we change the set of allowable transformations, the entire subject changes.

To summarize: The set of allowable transformations does not just define what it means for two figures to be congruent, it also defines what it means for a property to be "geometric." Position and orientation of a triangle, for example, are not geometric properties because neither property is preserved under the set of Euclidean transformations. By contrast, the distance between two points of a figure is a geometric property because distances are preserved under any sequence of Euclidean transformations.

At some level, Euclid may have been aware of all these ideas, but if he was, they were probably less of a concern for him than they are to contemporary mathematicians. The reason is that in Euclid's time there was only one set of axioms and one set of allowable transformations. Today, there are many geometries. Each geometry has its own vocabulary, its own axioms, and its own set of allowable transformations. Some of these geometries have proven to be good models for some aspects of physical space, but some of these geometries have no apparent counterpart in the physical world. From a modern point of view, this hardly matters. What is important is that the axioms are logically consistent and independent, and the class of allowable transformations is specified.

Topological systems are also axiomatic systems, and the notion of a topological transformation is just as central to a topological system as the notion of a Euclidean transformation is to Euclidean geometry. In this sense, topology can be described as a "close relative" of geometry. A crucial difference between topology and geometry lies in the set of allowable transformations. In topology, the set of allowable transformations is much larger and conceptually much richer than is the set of Euclidean transformations. All Euclidean transformations are topological transformations, but most topological transformations are not Euclidean. Similarly, the sets of transformations that define other geometries are also topological transformations, but many topological transformations have no counterpart in these geometries. It is in this sense that topology is a generalization of geometry.

What distinguishes topological transformations from geometric ones is that topological transformations are more "primitive." They retain only the most basic properties of the sets of points on which they act. In Euclidean geometry, when two figures are equal, they look the same. They have the same shape and the same size. Equality in Euclidean geometry is a very strict criterion. By contrast, in topology, two point sets that are topologically the same will often appear superficially to be very different. Our visual imagination is often of little use in determining topological equivalence.

## The Beginnings of Calculus

To appreciate some of the motivations for creating topology, it helps to understand some of the logical shortcomings in early conceptions of calculus. The German mathematician Gottfried Leibniz (1646–1716) created calculus. (Isaac Newton discovered calculus independently of Leibniz, and his formulation of calculus had the same sorts of logical shortcomings as those of Leibniz. Only Leibniz's contribution is considered here, because in addition to calculus, Leibniz also speculated on the possibilities inherent in a geometry that was independent of the idea of measurement. He called it *analysis situs*, and analysis situs became the forerunner of topology.)

Leibniz spoke several languages. In addition, he was a legal scholar, a philosopher, a diplomat, an inventor, a scientist, and a mathematician. He is best remembered, as mentioned above, as the codiscoverer of calculus, but he also made many other mathematical discoveries, including the binary number system, which had no apparent use until the invention of the computer. He invented one of the earliest mechanical calculators, and he was a writer who simultaneously maintained lively correspondences with hundreds of people. He is sometimes described as the last completely educated European in the sense that he is often credited with being the last person to attain a high level of expertise in every academic discipline known during his time.

Leibniz expressed his ideas about calculus in the language of geometry. One problem to which he gave a great deal of thought involved finding the slope of the line tangent to a given curve at a given point. (Recall that a line that is tangent to a curve at a particular point is the best straight line approximation to the curve at that point.) See the accompanying diagram. The problem of finding a tangent may sound too abstract to be useful, but historically it is one of the most important problems in mathematics. In fact,

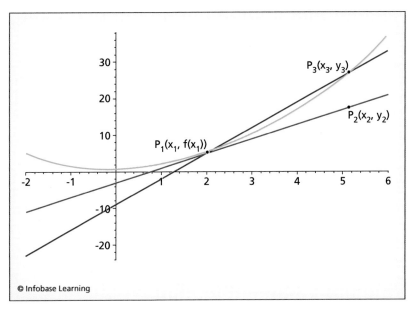

© Infobase Learning

*The red line is a secant line, and the blue line is the line tangent to the curve at the point P₁. The tangent is determined by, for example, P₁ and P₂, but the coordinates of P₂ are unknown. If P₃ is used instead, then, as the distance between P₃ and P₁ approaches zero, the slope of the secant line approaches the slope of the tangent line.*

the problem and its solution lie at the very heart of calculus, and the ability to compute the tangent has many applications to problems in engineering and science as well as mathematics.

As is indicated in Euclid's first postulate, a line is determined when any two points on the line are known. If Leibniz could find two points on the tangent line—say and $P_1(x_1, y_1)$ and $P_2(x_2, y_2)$—he would also know the slope of the line according to the following formula:

$$(2.1) \; slope = \frac{y_2 - y_1}{x_2 - x_1}$$

(Here $P_1$ and $P_2$ are the names of the points on the tangent line, and $(x_1, y_1)$ and $(x_2, y_2)$ are the coordinates of the points $P_1$ and $P_2$, respectively. Formula (2.1) is discussed at length in every high school algebra and trigonometry class.)

Leibniz's problem was that he only knew $P_1$, the point on the curve through which the tangent passes. He did not know $P_2$, nor did he have a way of finding $P_2$, and he could not determine the tangent from knowledge of $P_1$ alone. He could, however, *approximate* the unknown tangent by choosing another point on the curve—in the diagram that point is labeled $P_3(x_3, y_3)$—and by employing formula (2.1) with $(x_3, y_3)$ written in place of $(x_2, y_2)$. From $P_3$ he could compute an approximation to the slope of the tangent line. (The line determined by $P_1$ and $P_3$ is called a secant line, or secant for short.) The quality of the approximation one obtains by using $P_1$ and $P_3$ instead of $P_1$ and $P_2$ is determined by the choice of $P_3$. If $P_3$ is close to $P_1$, the difference between the slope of the secant and the slope of the tangent will be small, and the closer $P_3$ is to $P_1$, the smaller the difference will be.

It might be tempting to think that because the slope of the secant approaches the slope of the tangent as $P_3$ approaches $P_1$ along the curve, the slope of the secant *equals* the slope of the tangent when $P_3$ finally "gets to" $P_1$. But this idea cannot be right. When $P_3$ coincides with $P_1$, the secant fails to exist. This is reflected in the formula for the slope: When $P_3$ coincides with $P_1$—and $(x_3, y_3)$ coincides with $(x_1, y_1)$—the formula for the slope becomes meaningless since the denominator is zero, and division by zero has no meaning. As a general rule, there is no choice for $P_3$ on the curve that will yield the slope of the tangent.

It would be hard to overstate the mathematical problems caused by these simple-sounding observations. In attempting to overcome these mathematical difficulties, Leibniz postulated the existence of a class of numbers that he sometimes called infinitesimals; other times he called them differentials. He imagined them as being greater than zero but smaller than any positive number. They were so small that no matter how many were added together, their sum was still less than any positive number. He often tried to explain the concept of differentials by way of analogy. He wrote, ". . . the differential of a quantity can be thought of as bearing to the quantity itself a relationship analogous to that of a point to the earth or of earth's radius to that of the heavens."

Leibniz represented an infinitesimal in the $y$-direction with the symbol $dy$ and an infinitesimal in the $x$-direction with the symbol $dx$. By adding $dx$ to $x_1$ and $dy$ to $y_1$, Leibniz obtained the point $P_3(x_1 + dx, y_1 + dy)$, which he envisioned as a point on the curve very close to $P_1$—Leibniz used the phrase "infinitely close" to $P_1$. In fact, according to Leibniz, $P_1(x_1, y_1)$ and $P_3(x_1 + dx, y_1 + dy)$ were so close that the difference between the two points was "less than any given length." Leibniz described differentials in these words:

> ". . . these $dx$ and $dy$ are taken to be infinitely small, or the two points on the curve are understood to be at a distance apart less than any given length . . ."

In a sense, $P_3(x_1 + dx, y_1 + dy)$ is the point on the curve that was immediately adjacent to $P_1(x_1, y_1)$. It is, in effect, the next point over.

Using formula (2.1) to compute the slope determined by $P_1(x_1, y_1)$ and $P_3(x_1 + dx, y_1 + dy)$ gives

$$(2.2) \; slope = \frac{y_1 + dy - y_1}{x_1 + dx - x_1}$$

or, upon simplification

$$(2.3) \; slope = \frac{dy}{dx}$$

(The algorithms Leibniz used to manipulate his infinitesimals in order to obtain numerical answers are omitted here.) The slope of the tangent line at $P_1(x_1, y_1)$ is better known as the derivative of the curve at $P_1(x_1, y_1)$. The derivative is a function. Its value at the point $x_1$ *is* the slope of the tangent line passing through $(x_1, y_1)$.

Keep in mind that according to Leibniz, $dx$ is not zero. Consequently, the denominator of the fraction in equation (2.3) makes mathematical sense (provided one is willing to accept the existence of infinitely small nonzero quantities), and the numerator in (2.3) will have approximately the same magnitude as the denominator, so their ratio "makes sense." Consequently, the value of the ratio in equation (2.3), which is the value of Leibniz's derivative, is determined by the relative sizes of the two infinitesi-

## COUNTEREXAMPLE 1: A CONTINUOUS FUNCTION THAT IS NOT EVERYWHERE DIFFERENTIABLE

The mathematicians of Leibniz's time took it for granted that one could find a tangent line at every point of a curve. Given a curve and a point $P_1$ on the curve, the tangent line can be constructed by passing a line through $P_1$ and another point $P_3$ lying on the curve. Any point $P_3$ different from $P_1$ will yield a line because any two points determine a line. To obtain the tangent, according to Leibniz, just allow $P_3$ to move "near enough" to $P_1$. The result, Leibniz asserted, had to be the tangent. This idea was formulated in terms of a general principle, which is now known as Leibniz's principle of continuity:

"In any supposed transition, ending in a terminus, it is permissible to institute a general reasoning, in which the final terminus may also be included."

However, this is false, as the following counterexample demonstrates.

Consider the function $f(x) = |x|$, where the symbol $|x|$ means the "absolute value of $x$." (As a matter of definition, $|x| = x$ if $x \geq 0$, and $|x| = -x$ if $x < 0$.) As is indicated in the accompanying diagram, the graph of this function lies in the first and second quadrants of the plane. The graph of $f(x)$ coincides with the graph of the line $y = x$ in the first quadrant, and in the second quadrant, it coincides with the graph of the line $y = -x$.

For each positive value of $x$, the tangent to the graph exists and coincides with the line $y = x$, and for each negative value of $x$, the tangent to the graph exists and coincides with the line $y = -x$. According to Leibniz's principle of continuity, it should be possible to extend to the origin the process of forming the tangent. The origin would be the "terminus," but if the origin is approached from the right, the tangent at the origin must have a slope coinciding with the line $y = x$–that is, the slope must be $+1$. If the origin is approached from the left, the slope of the tangent at the origin must coincide with the line $y = -x$–that is, the slope must be $-1$. The tangent at the origin is, therefore, impossible to define since it cannot simultaneously have a slope of $+1$ and $-1$. Leibniz's principle of continuity fails. There are points on some curves where the derivative fails to exist.

mals. Finally, because $P_1(x_1, y_1)$ and $P_3(x_1 + dx, y_1 + dy)$ are "infinitely close" together, the slope determined by these two points must, according to Leibniz, be the slope of the tangent. Why?

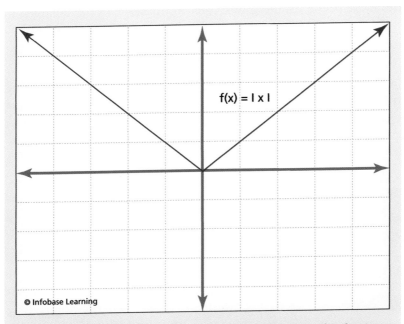

$f(x) = |x|$

© Infobase Learning

*The derivative is equal to +1 at every point on the curve that lies in the first quadrant, and the derivative is equal to –1 at every point on the curve that lies in the second quadrant. At the origin, therefore, the derivative does not exist.*

Mathematicians soon discovered many examples of functions with the occasional corner or cusp in their graph. At these points, the derivative failed to exist. The existence of such curves became geometrically "obvious" to them (as they become obvious to us) after a little thought. These mathematicians also believed that it was "obvious" that points where the derivative fails to exist are exceptional, in the sense that cusps and corners are "always" isolated from each other, and therein lies the problem with geometric reasoning. While it is hard to imagine a curve that is so "jagged" that it consists entirely of cusps and corners, such curves are more the rule than the exception. Geometric reasoning failed to reveal the existence of continuous nowhere *differentiable* functions because most of us cannot imagine what such jagged curves look like. Certainly their graphs cannot be drawn. Topological ideas would be required before calculus could be put on a firm logical foundation.

The (positive) difference between the slope of the secant and the slope of the tangent must be less than any given number because the two points are at a distance apart that is "less than any given length." This was the key: By choosing $P_3$ "infinitely close" to $P_1$, the slope of the resulting line equaled the slope of the tangent.

For Leibniz, infinitesimals were a convenient quasi-mathematical idea that enabled him to give expression to the very important mathematical idea of *closeness*. Calculus, as envisioned by Leibniz (and Newton), was not a sequence of logical deductions beginning from a set of axioms and definitions. It was not mathematics in the sense that the Greeks understood the word. Calculus was originally a set of techniques the justification for which was the results that were obtained by using them. Calculus made a huge difference in the progress of science and mathematics right from the time of its introduction, but even early in the development of calculus its logical shortcomings were apparent to many. The British philosopher and theologian Bishop George Berkeley (1685–1753) famously criticized the foundations of calculus when he wrote, "I say that in every other science men prove their conclusions by their principles and not their principles by their conclusions." Unfortunately, it is often easier to recognize an error than to fix one.

*Bishop George Berkeley. Calculus was originally expressed as a collection of algorithms unsupported by careful mathematical reasoning. Berkeley famously mocked the mathematicians of his time for "proving . . . their principles by their conclusions."* (National Portrait Gallery)

Without a logical framework to support his algorithms, Leibniz was unable to rigorously test his ideas. In retrospect, some of his ideas were correct, some were incorrect, and in some cases it is not quite clear what he meant. He was not alone. A rigorous understanding of the mathematical idea of closeness—an understanding that does not make use of infinitesimals—eluded mathematicians until well into the 19th century. By the latter half of the 19th century, mathematicians were forced to develop new ideas because the old ad hoc justifications proposed by Leibniz, Newton, and their successors, ideas that had once spurred progress in mathematics and science, had become a hindrance to further progress.

# 2

# A FAILURE OF INTUITION

For millennia, Greek geometry occupied a central place in the mathematical traditions of many cultures. Early mathematicians—some lived in present-day Turkey, present-day Iran and Iraq, North Africa, and southern Europe—regarded Greek geometry with a sort of reverence. They believed that there was little they could add to the study of geometry because the Greeks had already accomplished the better part of what could be done. The Greeks had, of course, left some problems unsolved, and later generations of mathematicians addressed themselves to these problems. Even so, for more than 2,000 years, mathematicians could not imagine a geometry other than the geometry of Euclid. They could not imagine another model for space than the one described in Euclid's *Elements*. Part of this chapter describes how, early in the 19th century, this concept of geometry was finally abandoned, and mathematicians began to consider so-called non-Euclidean geometries. For these mathematical pioneers, the new geometries did not have the same intuitive appeal as did Euclid's. In fact, the first mathematician to publish a description of a non-Euclidean geometry described his discovery as an "imaginary" geometry to distinguish it from what he perceived as the real geometry of the ancient Greeks. Nevertheless, the willingness to propose alternative sets of geometric axioms was an enormous conceptual breakthrough that prepared the way for early topologists, who often spent a great deal of time tinkering with alternative sets of topological axioms in order to generate "spaces" with various properties.

The second part of this chapter concerns an important attempt to correct the logical shortcomings in Leibniz's (and Newton's) intuitive conceptions of calculus. The substitution of logic for pictures is called the arithmetization of analysis. (Analysis is the branch of mathematics that grew out of calculus.) The drive to place analysis on a firm logical foundation also began in earnest in the early decades of the 19th century. As described in chapter 1, calculus was initially justified in two ways: first, by appealing to one's intuition about the existence of various limits and second by pointing to the important and often correct results obtained by using calculus algorithms. The problem is that in calculus, as in the rest of mathematics, intuition is often a poor guide to mathematical truth. By the late 18th century, mathematicians' intuitions were increasingly leading them astray. Putting calculus on a firm logical foundation became a matter of some urgency, and many of the best mathematicians of the 19th century applied themselves to this task. The arithmetization of analysis was one of the main motivations for early topological research.

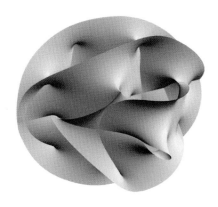

*Mathematics allows one to create infinitely many distinct shapes—most too complex to represent pictorially. By the latter half of the 19th century, it had become clear to many mathematicians that geometrically plausible arguments had no place within analysis.*

## An Alternative to Euclid's Axioms

In the centuries after Euclid, many mathematicians tried to show that Euclid had made a fundamental error: They believed that the fifth postulate was not a postulate at all. They believed that the fifth postulate could be proved as a logical consequence of the other postulates and axioms. In other words, they thought that Euclid's collection of postulates and axioms was not logically independent and that the fifth postulate, in particular, could be proved as a theorem, a necessary consequence of the other five axioms and four postulates.

In more modern terminology, the fifth postulate can be rephrased in the following way: "Given a line and a point not on the line, there exists exactly one line through the given point and parallel to the given line." Expressed in this way, it seems "obviously true," which is why mathematicians kept trying to prove it, and centuries of failure seemed only to inspire them to greater efforts. After all, just because previous generations of mathematicians had failed to prove the logical dependence of Euclid's axioms and postulates did not mean that it was impossible to do so. The proof may just have been exceptionally difficult (as opposed to impossible), and so for the next 2,000 years mathematicians struggled with the idea of proving the fifth postulate. They kept trying, in effect, to prove that Euclid had made a mistake.

Early in the 19th century, two mathematicians published papers that showed that Euclid had gotten it right all along—their methods of proof were similar, and neither was aware of the work of the other. As with calculus, this was another case of independent discovery. The mathematicians were the Russian Nikolay Ivanovich Lobachevsky (1792–1856) and the Hungarian János Bolyai. (The German mathematician Carl Friedrich Gauss and the German professor of law Ferdinand Karl Schweikart also had similar ideas. Neither published their ideas. Gauss was afraid of ridicule, and for reasons that are not clear, Schweikart also did not make his ideas widely known.)

Lobachevsky, who is sometimes called the "Copernicus of geometry," was from a family of modest means. He attended secondary school and the University of Kazan with the help of scholarships. He eventually found a position teaching at the University of Kazan and later worked as an administrator at his alma mater. Lobachevsky sought to improve his university and make it as accessible as possible. He seems to have given as much attention to his administrative duties as he did to his mathematical research. He was successful in both areas, but his work in mathematics changed the history of the subject.

Bolyai was the son of the prominent mathematician Farkas Bolyai, who had spent many long hours investigating Euclid's fifth postulate. He apparently considered his time with the fifth postulate poorly spent, and he pleaded with his son to find something else to do. Perhaps his son listened. Certainly mathematics was to János Bolyai only one of several interests. He was also a virtuoso violin player and a renowned swordsman. It seems that there were few things he could not do. What makes him important to the history of mathematics is that when he turned his attention to the fifth postulate, he was able to solve a 2,000-year-old puzzle.

Here is the method undertaken by both Lobachevsky and Bolyai: Rather than attack the problem directly and attempt to prove that the fifth postulate is (or is not) a logical consequence of the remaining axioms and postulates, they substituted an entirely different postulate *in place of* the fifth postulate. Although they used slightly different versions of the alternate postulate, the idea is the same. Here is a paraphrased version of their alternative to the fifth postulate: Given a line and a point not on the line, there exists more than one line passing through the point and parallel to the given line.

For most people, it is hard—or impossible—to visualize the situation described by Lobachevsky's and Bolyai's alternative to the fifth postulate. Keep in mind that this is planar geometry, and for most people there does not seem to be enough "room" in the plane to allow more than one line to pass through the given point and be parallel to the given line. Even Lobachevsky called the resulting geometry "imaginary," but the motivation behind the postulate is brilliant. If Euclid's version of the fifth postulate is actually a logical consequence of the other nine postulates and axioms, then substituting the alternative version creates a logical contradiction. Why? If Euclid's system of axioms and postulates were logically dependent, then both versions of the fifth postulate would be present in the geometry. The alternative version would be explicitly present, and Euclid's version would be present as a logical consequence of the other five axioms and four original postulates. It should, therefore, be possible to find a statement in Lobachevsky's and Bolyai's system that can be proved both true and false, but if Euclid was right and his version of the fifth postulate is actually logically independent of the other axioms and postulates, then changing it will not produce any logical contradictions. Instead, a new and logically consistent geometry will have been created. The resulting theorems would, however, be very different from those found in *Elements*.

What Lobachevsky and Bolyai discovered is that Euclid had been correct when he included the fifth postulate. The geometry that resulted by replacing Euclid's fifth postulate with the alternate version was entirely self-consistent. Because no logical contradictions arose, Euclid's fifth postulate could not be a logical consequence of the other postulates and axioms. To get a feel for the sorts of theorems that were proved from the alternative set of axioms, we have included two "elementary" theorems of the geometry of Lobachevsky and Bolyai together with the corresponding theorem in *Elements*.

1. Bolyai and Lobachevsky's geometry: The sum of the measures of the interior angles of every triangle is less than 180°. Euclid's geometry: The sum of the measures of the interior angles of every triangle equals 180°.

2. Bolyai and Lobachevsky's geometry: Two triangles are congruent whenever the pairs of corresponding angles are equal. (In other words, two triangles that are of the same shape must also be the same size.) Euclid's geometry: Two triangles are similar—but not necessarily congruent—whenever the pairs of corresponding angles are equal.

It took decades for mathematicians to become accustomed to the much more abstract approach of Bolyai and Lobachevsky, but eventually they came to accept the idea that all that was really required of a set of axioms was that they be logically consistent and logically independent. For example, if one wanted a particular property to be present in a geometry, one need only choose a set of axioms such that the property is mentioned explicitly in the axioms or such that the property is a logical consequence of the axioms. (In the 20th century, discoveries in logic proved that the situation with respect to the axiomatic method is more complex than is described here, but the axiomatic method remains fundamental to all mathematical research, and an appreciation of the power of the method really begins with the work of Bolyai and Lobachevsky.)

Early in the 20th century, topologists spent a great deal of effort testing different sets of axioms in order to determine which sets are logically equivalent and which sets produce nonidentical topological systems. The axioms are, not surprisingly, very different from those of Euclid and from those of Bolyai and Lobachevsky, but the perception of axioms as collections of abstract sentences that need only satisfy the conditions of consistency, independence, and completeness dates to the work of Bolyai and Lobachevsky. (The transformation group that characterizes the geometry of Bolyai and Lobachevsky would not be discovered until late in the 19th century. It is fairly technical and is not described here.)

## Bernhard Bolzano and Further Limitations on Geometric Reasoning

Bernhard Bolzano (1781–1848) was a citizen of the Austro-Hungarian Empire in what is now the Czech Republic. He was a priest who was also interested in mathematics and philosophy. Bolzano joined the faculty at the University of Prague in 1805. It was a time in European history when militarism was glorified, and wars of conquest were common. Bolzano was outspoken in his objections to both war and militarism, and in 1819, he was forced out of his position at the university because of his beliefs. Today, Bolzano is remembered as a brilliant mathematician whose work was often far ahead of that of his contemporaries. His work did not much alter the history of mathematics, however, because Bolzano did not publish very often. Many of his discoveries became known only many years after his death and after others rediscovered them. With respect to the history of topology, Bolzano was the first to imagine three very important concepts.

One of the problems that Bolzano considered was the meaning of continuity. Today, students begin to use the phrase *a continuous function* in junior high school, but no real attempt is made to

*Bernhard Bolzano. One of the most forward-thinking mathematicians in the history of mathematics, Bolzano had little influence on the development of the subject because he published few of his ideas.* (Dibner Library of the History of Science and Technology, Smithsonian Institution)

define the idea. Informally, a function is often said to be continuous if the graph of the function can be drawn without lifting one's pencil from the paper. This is probably the kind of definition that would have appealed to Leibniz and Newton as well. They depended on their geometric intuition to determine whether a function was continuous, and their intuition led them to consider only functions that were continuous and *differentiable*. (A function is differentiable if it has a derivative at each point in its domain.) As a consequence, they confused the concept of continuity with that of differentiability.

Continuity and differentiability are very different ideas, but for the kinds of functions that are considered in elementary algebra and geometry classes—and even in many calculus classes—the "pencil definition" is still all that is necessary. The reason is that elementary functions have graphs that are easily drawn, and with respect to the elementary continuous functions, one *can* draw their graphs without lifting one's pencil from the paper. Problems arise, however, when considering the class of *all* continuous functions. In most cases drawings are not in any way relevant to the determination of continuity.

What, then, does it mean for a function to be continuous? For Bolzano, continuity depended on the idea of "closeness," in the sense that small changes in the domain of the function caused small changes in the range, but making this idea precise requires some linguistic gymnastics. Bolzano's definition is equivalent to the following definition, which is expressed in more modern notation: Let the letter $f$ denote a function, and let $x_1$ represent a point in the domain of $f$. The function $f$ is said to be continuous at $x_1$ if for any given positive number $\varepsilon$ (epsilon) there exists a positive number $\delta$ (delta) with the property that whenever a point $x_2$ of the domain is within $\delta$ units of $x_1$, $f(x_2)$ is within $\varepsilon$ units of $f(x_1)$.

Here is another way of thinking about this definition. Imagine two intervals. One interval is centered at $f(x_1)$ and extends $\varepsilon$ units in either direction. If $f$ is continuous at $x_1$, there is another interval—this one centered at $x_1$ and extending $\delta$ units in either direction—such that if a point $x_2$ lies within the $\delta$-interval centered at $x_1$, then $f(x_2)$ lies in the $\varepsilon$-interval centered at $f(x_1)$. This is hardly an "intuitively obvious" definition! See the accompanying diagram.

Here is how the definition of continuity expresses the idea of "closeness:" The letter $\varepsilon$ represents the idea of closeness in the

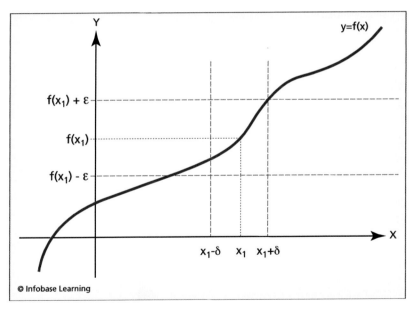

© Infobase Learning

*If $x_2$ is any point within the interval of width $2\delta$ centered at the point $x_1$ on the x-axis, then $f(x_2)$ lies within the interval of width $2\varepsilon$ centered at the point $f(x_1)$ on the y-axis. If for every value of $\varepsilon$ there exists a value for $\delta$ such that this condition is satisfied then the function $f$ is said to be continuous at $x_1$.*

range of $f$. It can be chosen arbitrarily small (as long as it remains greater than zero). The symbol $\delta$ represents the idea of closeness in the domain. Suppose we are given $\varepsilon$. If $f$ is continuous at $x_1$, then there is a $\delta$ such that whenever $x_1$ and $x_2$ are close—that is, whenever they are within $\delta$ units of each other—$f(x_2)$ will be close (within $\varepsilon$ units) to $f(x_1)$. The function $f$ is a continuous function if it is continuous at each point in its domain.

There are two important things to notice about this definition. First, $\delta$ and $\varepsilon$ can be, and often are, different in size. To see this, consider the function $f(x) = 1,000,000x$. The graph of this function is a line passing through the origin with slope 1,000,000. If $\varepsilon$ is 1, then any value of $\delta$ that is less than or equal to 0.000001 will satisfy the definition of continuity. (Because the graph of this function is a straight line, $\delta$ is the same for every value of $x$.) Second, as a general rule, the value of $\delta$ will usually depend on $x$ as well as $\varepsilon$. All we can say for sure is that if the function $f$ is continuous, then for every value of $x$ and every positive value of $\varepsilon$, some value of $\delta$ exists that satisfies the definition of continuity.

By now it is easy to see why the do-not-pick-up-the-pencil-off-the-paper definition of continuity is so much more widely used than Bolzano's definition, which is both hard to state and hard to appreciate. So why did Bolzano bother to develop it? (It is an interesting fact that a similar definition of continuity was developed at about the same time by the French mathematician Augustin-Louis Cauchy [1789–1857]. This is still another case of simultaneous discovery in mathematics, and Cauchy and Bolzano were similar in other ways. Cauchy was also a man of conscience, and he was punished for his decisions of conscience just as Bolzano was. In 1830, when Louis-Phillipe became king of France by deposing his predecessor, the Academy of Sciences, which was where Cauchy worked, instituted a loyalty oath as a condition of employment. All faculty were required to swear an oath to the new king. Cauchy refused. He left his position at the university rather than submit. He found work elsewhere in Europe, and for a time he worked in Prague, which was also where Bolzano was living. There is no evidence that the two of them ever met. Eight years later, Cauchy was able to return to the Academy of Sciences in Paris without having to swear an oath of allegiance. Initially, he was not permitted to teach, but eventually the institution relented, and Cauchy, one of the most productive and creative mathematicians in history, was permitted to return to his teaching duties as well.)

One reason that Bolzano's (and Cauchy's) definition of continuity is so important is that the function concept is so general. Mathematicians have found ways to create functions that are both continuous and impossible to draw. In fact, Bolzano made the first such function. Recall that mathematicians of the 18th century took it for granted that functions had derivatives everywhere—or at least everywhere except at exceptional points (see the sidebar "Counterexample 1: A Continuous Function That Is Not Everywhere Differentiable"). Bolzano described a function that was continuous at every point in its domain but had no derivative anywhere. Essentially, Bolzano's function has a corner between any two points on the curve no matter how close together the points are chosen. One way of thinking about Bolzano's function is, therefore, that it consists entirely of corners. (See the sidebar "Counterexample 2: A Continuous Nowhere Differentiable Function.")

Bolzano's nowhere differentiable curve was the first of its kind. It demonstrates the necessity for an abstract approach to the idea of continuity, just as it also demonstrates the limitations of geometric methods. Our "visual imagination" is of little value in understanding Bolzano's curve, and the don't-pick-up-the-pencil-off-the-paper criterion fails because the curve cannot be drawn. This explains why the movement to replace geometric reasoning in analysis with the abstract approach pioneered by Bolzano became a high priority for many mathematicians of the late 19th century. To make additional progress, mathematicians had to find a more precise and more productive way of understanding sets of points. This was one of the principal motivations for the development of topology. With respect to the arithmetization of analysis, Bolzano was the first.

Bolzano's nowhere differentiable curve had little impact on his contemporaries. The existence of such a curve would not be widely known until the latter part of the 19th century, when the

German mathematician Karl Weierstrass (1815–97) produced a formula for such a curve. The latter decades of the 19th century were also a time when mathematicians were more ready to accept the existence of what they called "pathological functions." As already mentioned, part of the reason that Bolzano's curve had less impact than Weierstrass's is that Bolzano's work was not widely known, but a second reason that his curve had less impact is that he did not give a "mathematical" formula for the curve. During Bolzano's time, mathematicians considered geometry to be richer than algebra, in the sense that they believed that they could draw many curves for which there was no corresponding algebraic function, but that given a function, one could always draw its graph. We now know that both assertions are false, but in their view Bolzano produced a curve, not a function.

A curve that consists entirely of corners still strikes many people as a paradoxical result. Bolzano enjoyed producing paradoxical results. Many of his paradoxes turned on the idea of *infinite sets* and infinite processes. He even wrote a book entitled *Paradoxes of Infinity*. Surprising people with the paradoxes of the infinite is not hard. Although most people assume that they know what it means to say that a set has infinitely many elements, most people, in fact, do not. The logical implications of the infinite are often difficult to appreciate.

Bolzano enjoyed demonstrating that sets that seemed to be very different in size were really the same size. Because these ideas are also important in the development of topology, we repeat a few of them here. Consider the intervals $\{x: 0 \leq x \leq 2\}$ and $\{x: 0 \leq x \leq n\}$, where $n$ represents any integer greater than 2. Because $\{x: 0 \leq x \leq 2\}$ is a proper subset of $\{x: 0 \leq x \leq n\}$, it might seem that the shorter interval has fewer elements than the longer one, but no matter how large we choose $n$, we can prove that the two sets are exactly the same size. Here is the proof: Consider the function $f(x) = \frac{nx}{2}$ with domain $\{x: 0 \leq x \leq 2\}$. The range of this function is $\{y: 0 \leq y \leq n\}$. (Its graph is a line segment.) For each $x$ in the domain, there is exactly one element in the range with which it is paired, and for each element in the range, there is exactly one element in the domain with which it is paired. In other words, there is a one-to-one relation-

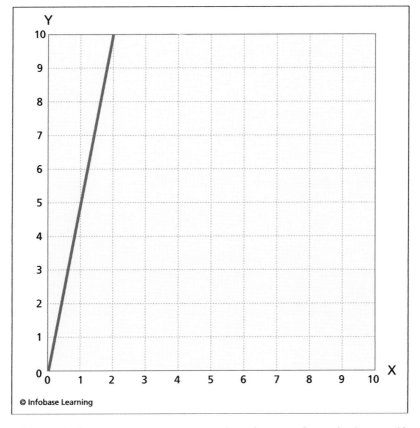

*This graph shows a one-to-one correspondence between the set {x: 0 ≤ x ≤ 2} and the set {y: 0 ≤ y ≤ 10}, thereby demonstrating that there are as many points in the short interval as there are in the long one.*

ship between the points in the domain and the points in the range even though the domain is a proper subset of the range. (See the accompanying diagram.) This "paradox" is now known to be the defining property of an infinite set: A set is infinite if it can be placed into *one-to-one correspondence* with a proper subset of itself.

Bolzano was not the first to write about this property of infinite sets. He was the second. The first was the Italian scientist and mathematician Galileo Galilei (1564–1642), who noticed that the set of all natural numbers could be placed in one-to-one correspondence with the set of all perfect squares. Galileo's description was a bit wordy because good algebraic notation had not been invented yet, but in modern notation he noticed that he could define a function on the natural numbers according to the following formula:

## COUNTEREXAMPLE 2: A CONTINUOUS NOWHERE DIFFERENTIABLE FUNCTION

The 18th-century belief that the points at which a function fails to have a derivative are isolated from one another—and in that sense, "exceptional"—is false. The first counterexample was produced by Bolzano. Here is his method for producing a curve that is continuous at each point but fails to have a derivative at any of its points. His counterexample dates to 1830.

Step 1: Draw a nonhorizontal line segment $PQ$ and label the midpoint of the segment $M$.

Step 2: Draw a horizontal line though $M$, and draw a second horizontal line through $Q$. Find the point on the segment $PM$ that is three-fourths the distance from $P$ to $M$, and label it $x$. Find the point on the segment $MQ$ that is three-fourths the distance from $M$ to $Q$, and label it $y$.

Step 3: Reflect the point $x$ about the horizontal line passing through $M$. Label this new point $x'$. Reflect the point $y$ about the horizontal line passing through $Q$, and label this new point $y'$.

Step 4: Using straight line segments, connect $P$ to $x'$ and connect $x'$ to $M$. Similarly, use straight line segments to connect $M$ to $y'$ and $y'$ to $Q$. The result is a graph with four straight segments and three corners.

Now repeat steps 1 through 4 on each of the resulting segments.

Continue to repeat steps 1 through 4 on each of the resulting segments to produce a graph with an ever-increasing number of corners.

This produces a sequence of graphs of functions. The formulas for the functions in this sequence are not especially difficult to find, but the formulas are long and not especially informative. Instead of deriving the formulas, label the first function $f_1$. This is the straight line segment with which we began. Label the second function $f_2$. This function has the red graph. It represents the result at the end of step 4. If we repeat the procedure on each segment of the red graph, we get the green one. Call the function with this graph $f_3$. By continuing in this way, we obtain a sequence of such functions $f_1, f_2, f_3, \ldots$ The further one goes in the sequence, the more closely spaced the corners are placed on each graph. Now imagine vertical parallel lines.

$f(n) = n^2$. His function paired each natural number $n$ with a perfect square $n^2$: The number 1 is paired with 1; the number 2 is paired with 4; the number 3 is paired with 9; and, in general, the number

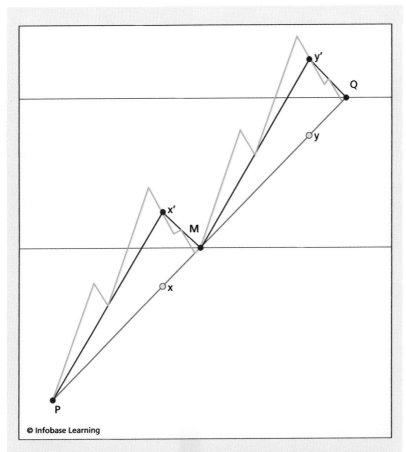

*Bolzano's procedure begins with the straight (blue) line. The red line is obtained at the completion of step 4 of the procedure. The green line is obtained after applying steps 1–4 to each of the straight segments that make up the red line. Repeat again and again. The resulting graph converges to one that is continuous and nowhere differentiable.*

No matter how close the lines are placed to one another, all functions with a large enough subscript will have at least one corner somewhere between those vertical lines. The sequence of functions described in the algorithm determines a "limit function" that has corners everywhere. (The proof of this last statement is too difficult to produce here.)

$n$ is paired with $n^2$. As Galileo pointed out, it might seem that there are far fewer perfect squares than natural numbers because the percentage of perfect squares in the set consisting of the first $n$ natural numbers approaches zero as $n$ becomes large, but because the set of natural numbers is infinite, a one-to-one correspondence can still be established.

Although he was careful to document his "paradoxical" results, Bolzano never used them to develop a clear understanding of the real number system. A clear conception of the real number system did not develop until the latter part of the 19th century, when mathematicians learned how to manipulate infinite sets. A deeper understanding of infinite sets enabled them to obtain a new and deeper understanding of mathematics in general. The mathematician most responsible for establishing the theory of sets is also generally credited with founding the modern era in mathematics. His work also marks the introduction of general topology, which he helped to create in order to overcome some of the problems described so far in this narrative.

# 3

# A NEW MATHEMATICAL LANDSCAPE

During the latter half of the 19th century, modern mathematics began to take shape. The geometric constructions of Leibniz and Newton—a visual language that had proven to be inadequate to describe subsequent discoveries—were finally replaced by a more abstract language founded on the theory of sets.

At first glance, sets are about as primitive a concept as can be imagined. The concept of a set, which is, after all, a collection of objects, might not appear to be a rich enough idea to support modern mathematics, but just the opposite proved to be true. The more that mathematicians studied sets, the more astonished they were at what they discovered, and *astonished* is the right word. The results that these mathematicians obtained were often controversial because they violated many common sense notions about equality and dimension. Some mathematicians were left by the wayside complaining about mathematical "absurdities." The more adaptable ones changed their notions of what constituted "common sense notions." Georg Cantor, who did more than anyone else in establishing the theory of sets, is said to have exclaimed about one particularly remarkable proof of his own making, "I see it but I don't believe it!"

As these mathematicians learned more, they began to impose additional structure on their sets. They began to distinguish among infinite sets of different sizes and different (topological) properties. Their research revealed an entirely new mathematical landscape full of exotic mathematical objects, but more than the answers they obtained, their research raised new questions. Ideas

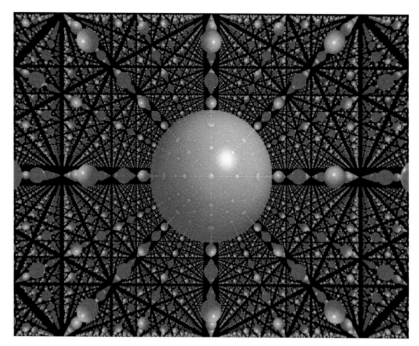

*Some of the logical implications of the concept of infinity strike many people as strange, even today. Part of what it means to learn mathematics is becoming accustomed to the strangeness of the infinite.* (Casey Uhrig)

as basic to science and mathematics as that of dimension were called into question. One of the accomplishments of set-theoretic topologists is that their discoveries helped to restore order to mathematics by identifying (or creating, depending on one's point of view) underlying patterns, the existence of which no one had previously suspected.

## Richard Dedekind and the Continuum

The German mathematician Richard Dedekind (1831–1916) made an essential step in arithmetizing analysis by making explicit the relationship between real numbers and the real number line. The ancient Greeks had assumed that a line is a *continuum* of points. They also assumed that, in contrast to the line, numbers did not form a continuum. These two ideas, the continuity of lines and the discrete nature of numbers, were accepted by mathematicians for the next 2,000 years. In his famous paper "Continuity and Irrational Numbers," Dedekind questioned both ideas. He wrote that the belief that the line forms a continuum was an assumption—one could not prove this statement—but *if* the line were continuous, so was the set of all real numbers. His conception of the real number line continues to influence mathematics on an elementary and advanced level today.

Dedekind studied mathematics at Göttingen University under Carl Friedrich Gauss, one of the leading mathematicians of the 19th century. For seven years, he taught at the university level, first at Göttingen and later at Zurich Polytechnic. For the next 50 years, he taught in Braunschweig, Germany, at the Technical High School, a remarkable choice for one of the most forward-thinking mathematicians of his age.

Since the time of the Greeks, irrational numbers had remained something of a puzzle. Recall that rational numbers are numbers that can be represented as the quotient of two whole numbers. The numbers ½ and ¾, for example, are rational numbers, but the number $\sqrt{2}$ is not rational because there is no choice of whole numbers, *a* and *b*, such that their quotient, *a/b*, when squared, equals 2. The number $\sqrt{2}$ is, therefore, an example of an irrational number. More generally, the set of irrational numbers is defined to be the set of numbers that are not rational.

Nevertheless, to defined something by what it is not yields very little information about what it is. The definition of irrational numbers as not rational goes back to the Greeks, who considered geometry and arithmetic to be very separate subjects, in part because geometry (as they understood it) dealt with continuously varying magnitudes—lines, surfaces, and volumes, for example—and arithmetic was concerned with numbers, which they regarded as discrete entities, but to do analysis rigorously, mathematicians needed a continuum of numbers. In other words, they needed as many numbers as there are points on a line. Dedekind established a *one-to-one correspondence* between the points on a line and the set of real numbers. He demonstrated that he could "pair up" points and numbers in the following one-to-one way: Each number was paired with exactly one point on the line, and each point on the line was paired with exactly one number. Through his correspondence, Dedekind demonstrated that the set of all real numbers forms a continuum provided one was willing to accept the continuity of the real line. (To be clear about the concept of one-to-one correspondence, imagine an auditorium in which everyone is seated. If every chair is taken, then there are as many people as chairs in the auditorium. If some of the chairs are empty, then there are more chairs than people. We may not know how many people or chairs are in the auditorium, but the concept of one-to-one correspondence still enables us to draw conclusions about the relative sizes of the set of people in the auditorium and the set of chairs in the auditorium.)

In Dedekind's model of the real number system, the rational numbers play a special role. These numbers can be placed into one-to-one correspondence with a subset of points on the real line. To see how, begin with a line. Identify one point on the line as zero, and identify a second point (to the right of zero) as the number one. Now measure off distances on the line corresponding to the natural numbers. They are located to the right of zero and are multiples of the zero-one distance. Once the points corresponding to the natural numbers have been located, use them to identify points corresponding to the rational numbers. (Any rational number can be expressed in terms of sums, differences, products, and quotients of natural numbers.) Positive rational numbers correspond to distances to the right of zero; negative rational numbers correspond to distances to the left of zero. We will call the points corresponding to these rational distances "rational points." However, this procedure leaves numerous "gaps" in the line. It does not, for example, generate a point corresponding to $\sqrt{2}$ or a point corresponding to the number $\pi$, because neither of these is a rational number. The question Dedekind sought to answer is, "What is the nature of these gaps?"

Dedekind expressed his answer in terms of cuts of the real line, now famously called *Dedekind cuts*. Each cut identifies a unique point, which is the location of the cut, and every point determines a unique cut. His method shows that there exist as many numbers as cuts. In particular, he shows that the set of all real numbers forms a continuum. Mathematicians now call this property "completeness," but Dedekind wrote that his construction demonstrated that the domain of real numbers ". . . had the same *continuity* as the straight line."

To establish Dedekind's correspondence imagine making a cut at a point $P$, where $P$ is some arbitrarily chosen point on the line. The point $P$ divides the set of rational points into two *disjoint* sets. (Imagine a string stretched horizontally from left to right. Cutting the string *partitions* the fibers of the string into two disjoint sets. Some fibers lie to the left of the cut and some to right. No fiber lies in both parts.) In just the same way, Dedekind's cut partitions the set of rational numbers into two sets, which we will call $A_1$ and $A_2$. Let $A_1$ be the set of rational numbers to the left of the cut, and let $A_2$ be the set of rational numbers to the right of the cut. Each number in $A_1$ is, therefore, less than each number in $A_2$. But what about $P$, the point at which the cut was made? Three possibilities exist: $P$ may belong to $A_1$; $P$ may belong to $A_2$; or $P$ may not belong to either $A_1$ or $A_2$.

If $P$ belongs to $A_1$, it is the largest element in $A_1$, and it is also a rational number because every element in $A_1$ is a rational number. Also, if $P$ belongs to $A_1$, then $P$ does not belong to $A_2$ because the sets share no points in common. Consequently, the set $A_2$ does not have a smallest number. To see why this is so, imagine that $A_2$ did have a smallest number. Call that number $Q$. Because we have already assumed that $P$ belongs to $A_1$ and that $A_1$ and $A_2$ have no elements in common, $Q$ must be bigger than $P$, but between any two rational numbers on the real line there is always a third rational number distinct from both. If the third number existed, it would not belong to either $A_1$ or $A_2$, since it would be larger than $P$ and smaller than $Q$. This contradicts the fact that every rational number belongs either to $A_1$ or $A_2$. We conclude, therefore, that if $P$ belongs to $A_1$, $A_2$ does not have a smallest element.

Now assume that $P$ belongs to $A_2$. Reasoning similar to that of the preceding paragraph shows that $A_1$ cannot have a largest element.

Finally, suppose that $P$ does not belong to either $A_1$ or $A_2$. Because every rational number belongs to either $A_1$ or $A_2$, $P$ must be an irrational number, but $P$ corresponds to the unique number that partitions the set of all rational numbers into the sets $A_1$ and $A_2$. The way that $P$ divides the rational numbers also reveals which irrational number corresponds to $P$. It is the unique number less than every rational number in $A_2$ and greater than every rational number in $A_1$. (Only one real number satisfies both criteria.) In this way, Dedekind showed that to each number corresponds exactly one point and to each point corresponds exactly one number. He concluded that the set of real numbers has the same continuity as the line. (Notice that Dedekind's correspondence allows us to use the terms *real number* and *point on the real line* interchangeably, and we will do exactly that in the next section.)

## Georg Cantor and Set Theory

The German mathematician Georg Cantor (1845–1918) did more than anyone else to place set theory at the center of mathematical thought. He was a creative mathematician with an unusual background. In university, he studied philosophy and physics in addition to mathematics. A mystic as well as a mathematician, he hoped his research into the nature of infinite sets would reveal something about the mind of God. The discoveries that Cantor made about infinite sets were so unexpected that mathematics journals were sometimes reluctant to publish his papers. Editors were worried that the papers contained errors, although they could find none. His research was so controversial that despite the fact that he changed the history of mathematics, he was refused a position that he wanted at the University of Berlin because of the sensational nature of his discoveries. It was characteristic of Cantor that he titled his Ph.D. thesis "In mathematics the art of asking questions is more valuable than solving problems."

Much of Cantor's work involved the problem of classifying infinite sets, and one of his main methods was the use of one-to-one correspondences. When he could prove that a one-to-one correspondence existed between two sets, he would conclude that the two sets were equal in size. When he could show that a one-to-one correspondence did not exist, he would conclude that

*This sculpture by Charles Perry is titled* Continuum. *Cantor showed that the continuum of real numbers cannot be put into one-to-one correspondence with the set of all rational numbers. In other words, infinite sets come in different sizes.* (Smithsonian National Air and Space Museum)

they were of different sizes. Cantor began by defining an infinite set as any set that can be put into a one-to-one correspondence with a proper subset of itself. (Recall that neither Galileo nor Bolzano attempted to be precise about the sizes of the sets that they considered. Instead, they simply demonstrated that a given infinite set could be placed in one-to-one correspondence with a proper subset of itself.)

Cantor soon demonstrated that he could place the set of natural numbers into a one-to-one correspondence with the set of all rational numbers. This was surprising to many people at the time, and many people are surprised by this result today. The reason is that between any two distinct real numbers—no matter how closely they are spaced—there are infinitely many rational numbers, and between any two distinct real numbers—no matter how far apart they are spaced—there are only finitely many natural numbers. Nevertheless, the set of rational numbers is no larger than the set of natural numbers, Cantor also discovered that there are even larger-looking sets than the set of rational numbers that can be placed in one-to-one correspondence with the set of natural numbers. The set of *algebraic numbers*, for example, is the set of all numbers that are roots of polynomials with rational coefficients. (Or to put it another way: An algebraic number is a solution to an equation of the form $a_n x^n + a_{n-1} x^{n-1} + \ldots + a_1 x + a_0 = 0$, where all the numbers $a_0, a_1, a_2, \ldots, a_n$ are rational numbers.) The set of algebraic numbers contains the set of rational numbers as a proper subset. It also contains many irrational numbers, but it, too, can be placed in one-to-one correspondence with the set of natural numbers. (Any set that can be placed in one-to-one correspondence with the set of natural numbers is called a *countable set*.)

When Galileo was faced with the "paradox of the infinite," he had speculated that the terms *less than, greater than,* and *equal to* did not apply to infinite sets, but Cantor showed that this speculation was false. While many different-looking sets are the same size, some sets cannot be placed in one-to-one correspondence with the set of natural numbers. Cantor showed, for example, that the set of real numbers is definitely larger than the set of natural numbers because the assumption that a one-to-one correspondence exists between these two sets leads to a logical contradiction.

To better appreciate what Cantor discovered, consider the so-called *power set*. Given a set $S$, the power set of $S$, which is often denoted by the symbol $2^S$, consists of all the subsets of $S$. To take a finite example, let $S = \{a, b, c\}$. The power set of $S$ is $\{\{a, b, c\}, \{a, b\},$

*{a, c}, {b, c}, {a}, {b}, {c}, ∅}*, where the symbol ∅ denotes the *empty set*, the set consisting of no elements. (The empty set is always assumed to be a subset of every set.) Notice that in our example, the set S has three elements, and the power set of S has eight elements, and $8 = 2^3$. For any finite set S, the number of elements in S is always $2^S$, which denotes the number 2 raised to the power of the number of elements in S. The notation is simply carried over to infinite sets. What Cantor discovered is that no matter whether the set is finite or infinite, the power set of a set can never be placed in one-to-one correspondence with the original set. In particular, the power set of the natural numbers is too large to be placed in one-to-one correspondence with the set of natural numbers. Instead, it can be placed in one-to-one correspondence with the set of all real numbers. Similarly, the set of all real numbers cannot be placed in one-to-one correspondence with the power set of the set of real numbers. The power set of the set of real numbers is, therefore, strictly larger than the set of real numbers.

To bring some order to his many discoveries, Cantor created a set of symbols to represent different "sizes" of infinity. Any set that could be placed in one-to-one correspondence with the natural numbers was said to have *cardinality* $\aleph_0$. (The symbol $\aleph$ is pronounced "aleph," and the word "cardinality" we can take to mean "number of elements.") Every countable set, therefore, has cardinality $\aleph_0$. Any set that can be placed in one-to-one correspondence with the power set of the natural numbers is said to have cardinality $\aleph_1$; any set that can be placed in one-to-one correspondence with the "power set of the power set" of the natural numbers is said to have cardinality $\aleph_2$, and so on. These symbols are called transfinite numbers, and Cantor even created a transfinite arithmetic. (Given two sets, A and B, suppose we know the cardinality of A and the cardinality of B, then Cantor's transfinite arithmetic enables us to compute the cardinality of the set formed by, for example, the *union* of A and B.)

What was more surprising to Cantor was that the dimension of a set seemed to play no role with respect to its size. As Cantor became increasingly adept at imagining one-to-one correspondences, he discovered a one-to-one correspondence between the points on the *unit interval*—this is the set {x: 0 ≤ x

≤ 1}—and the points in the *unit square*, which is the square with the unit interval as base—in other words, $\{(x, y): 0 \leq x \leq 1, 0 \leq y \leq 1\}$. (To be clear: The unit square is the *boundary* of the square and every point inside the boundary of the square.) The existence of such a correspondence was a shock. Cantor initially took it as "obvious" that the set of points in the unit square could not possibly be placed in one-to-one correspondence with the set of points on the unit interval. It seemed to Cantor, as it seems to many people today, that the dimensions of the two sets should affect the sizes of the sets. One can picture a unit square as consisting of infinitely many "vertical" copies of the unit interval, in the sense that above each point $x$ in the unit interval, there is (within the unit square) a unit interval with $x$ as an endpoint. In fact, the union of all such intervals is the unit square, and yet Cantor proved that there are no more points in the unit square than there are in the unit interval. (One way of thinking about this discovery is that it contradicts Euclid's fifth axiom, "The whole is greater than the part.")

Cantor went further, discovering a one-to-one correspondence between points in the unit interval and points in the *unit cube*, which is the term for a cube with edges one unit long. He even discovered a one-to-one correspondence between the unit interval and the unit "hypercube," which is a cube in $n$-dimensional space, where $n$ is any integer greater than three. He wrote a letter describing his discovery to his friend Richard Dedekind, and it is in that letter that he wrote, "I see it but I don't believe it!" His new results, Cantor believed, were calling into question fundamental ideas about the concept of dimension. What did it mean to say that a figure is $n$-dimensional when the points that constituted the figure could be placed in one-to-one correspondence with an $m$-dimensional figure, where $m$ is different from $n$?

Dedekind speculated that the key to understanding dimension required the concept of continuity. It was not enough, he believed, that the correspondence be one-to-one, it must also be continuous. In a letter to Cantor, Dedekind wrote that every one-to-one correspondence between sets of different dimensions must be discontinuous. This is easier said than proved, however, and neither Cantor nor Dedekind was able to prove Dedekind's hypothesis. The situation became cloudier still when the mathematician Giuseppe Peano discovered his continuous *space-filling curve* (See the sidebar "Peano's Space-Filling Curve").

Even so, infinite sets of points have other properties besides their cardinality, and there are things to discover besides correspondences. Cantor soon turned his attention to some of these other properties. Sets can, for example, be distinguished by the way they are distributed in "space." Compare, again, the set of natural numbers with the set of rational numbers. It is always possible to draw a circle containing a given natural number in its interior with the additional property that no other natural number belongs to the interior of the circle. In a sense, therefore, natural numbers are "isolated" from each other.

By way of contrast, if we draw a small circle centered about any point on the real line, that circle will always contain infinitely many rational numbers. This is true no matter how small we make the circle and no matter which point we choose as the center of the circle. (Dedekind made indirect use of this property when he showed that the set of real numbers has the same continuity as the line.) Although the set of natural numbers and the set of rational numbers have the same cardinality, the ways they are distributed along the line are very different.

Other possibilities exist. Consider, for example, the set $S$ = {1, ½, ⅓, . . .}. This set can also be placed in one-to-one correspondence with the set of natural numbers using the following function: $f(n) = 1/n$. As with the natural numbers, every number in the set $S$ is "isolated" from every other number in $S$ in the sense that no matter which element of $S$ we choose, we can draw a circle that contains that number in its interior and such that no other number in $S$ belongs to the interior of our circle. But the number zero, which does not belong to $S$, has a very different property: Every circle that contains zero in its interior contains infinitely many elements of the set $S$. The distribution of the elements of $S$ is very different from the distribution of the set of natural numbers, and it is also very different from the distribution of the set of rational numbers.

# COUNTEREXAMPLE 3: PEANO'S SPACE-FILLING CURVE

As mathematicians sought to phrase their mathematical ideas in ever more rigorous language, they discovered many "pathological functions," which is the name they gave to the functions that violated their common sense notions of what a function should be. The existence of pathological functions indicated that the logical implications of certain simply stated concepts could be both subtle and counterintuitive. One of the most famous of these pathological functions was discovered by the Italian mathematician and logician Giuseppe Peano (1858–1932). Roughly speaking, it provided a counterexample to the idea that the graph of a continuous function defined on an interval should "look like" a curve.

To appreciate Peano's curve, it is important to recall the way that functions are first introduced in an elementary algebra class. A function is defined as a set of ordered pairs with a particular property. Recall that the first element in the ordered pair belongs to a set called the domain, and the second element in the ordered pair belongs to a set called the range. In order that a set of ordered pairs determine a function, it is only

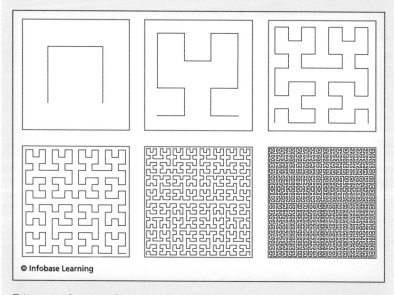

© Infobase Learning

*Diagram showing the process by which Peano's curve is generated. At each step, the square is divided into finer "cells," and then the pattern is repeated within each cell. In the limit, the process generates a curve that passes through each point in the unit square at least once.*

necessary that each element in the domain occur exactly once among the set of all ordered pairs that make up the function. The domain and the range can be sets of any type. In Peano's case, the domain is the unit interval $\{t: 0 \leq t \leq 1\}$, and the range is the unit square $\{(x, y): 0 \leq x \leq 1; 0 \leq y \leq 1\}$. (We use $t$ instead of $x$ in describing the unit interval just to avoid any possible confusion between the two sets.) Consequently, if we were to write one of the ordered pairs that belongs to Peano's function, it would look like this: $(t, (x, y))$. The number $t$ belongs to the domain, and the ordered pair $(x, y)$ belongs to the range. What drew the attention of researchers around the world to Peano's function—and it is a function according to our definition—is that it satisfied the following two criteria: (1) Every point in the unit square occurs at least once in the set of all ordered pairs that make up the function, and (2) the function is continuous. Geometrically speaking, therefore, Peano succeeded in continuously "deforming" the unit interval until it covered the unit square. (Peano's space-filling curve does not pass the so-called vertical line test, the test students learn in high school algebra courses, but the vertical line test applies only to functions in which the domain is a subset of the real numbers and the range is also a subset of the real numbers.)

Peano's curve called into question the concept of dimension. For a long time, mathematicians had naively accepted the idea that the dimension of the square was different from the dimension of the line *because* they had always used two numbers—the $x$ and $y$ coordinates—to identify a point in the square, and they only used one number to identify a point in the interval. However, Peano's curve could be interpreted as a scheme for identifying every point $(x, y)$ in the unit square with a single "coordinate," its $t$-coordinate, where $t$ is the point in the domain that is paired with $(x, y)$ by Peano's function, and unlike Cantor's correspondence between the interval and the square, Peano's curve is continuous.

Here are two additional facts about Peano's curve: (1) It passes through some points in the square more than once, so, in contrast to Cantor's correspondence, Peano's curve is not a one-to-one correspondence, and (2) the curve does not pass through any point in the square more than finitely many times. This raises the question of whether it can be refined so as to make it a continuous one-to-one correspondence.

Peano's curve has been studied intensively in the century since its discovery. It is now known that every continuous function with domain equal to the unit interval and range equal to the unit square must pass through some points of the square at least three times. Cantor's one-to-one correspondence between the unit square and the unit interval can, therefore, never be refined in such a way that it becomes a continuous one-to-one correspondence. Peano's discovery also inspired research into what would soon become a branch of set-theoretic topology called dimension theory. (Dimension theory is discussed in chapter 7.)

Faced with a rich collection of examples, Cantor began to search for unifying concepts. He developed a vocabulary that would enable him to describe the extraordinarily rich mathematical landscape that he had uncovered. A term (and a concept) that is of special importance to set-theoretic topology is the term *open set*. To understand the idea of an open set—and we will use the term often in what follows—suppose that we are given a set $S$ and a point $x$ that belongs to $S$. (To be specific, suppose that $S$ is a subset of the real line or of the plane.) The point $x$ is called an *interior point* of $S$ if we can draw a small circle centered about $x$ such that only points of $S$ belong to the interior of the circle. In other words, a point $x$ in $S$ is an interior point of $S$ provided that every point that is close to $x$ also belongs to $S$.

Consider, for example, this subset of the real line: $\{x: 0 < x < 1\}$. It is an example of an open set. To see why this is true, choose any point belonging to this set, and you will see that it is possible to draw a circle containing the point with the additional property that all points within the circle also belong to $\{x: 0 < x < 1\}$. By contrast, the set $\{x: 0 \leq x \leq 1\}$ is not open. Why? If we draw a circle about 1, for example, the circle will contain numbers that are greater than 1, and these numbers do not belong to $\{x: 0 \leq x \leq 1\}$. This is true no matter how small the circle is drawn. Similar reasoning shows that any circle, no matter how small, when drawn about 0, will contain numbers that are not elements of $\{x: 0 \leq x \leq 1\}$. (Notice, however, that the set of all real numbers that do *not* belong to $\{x: 0 \leq x \leq 1\}$ is an open set. This is, in fact, how a *closed set* is defined: If $S$ is an open set, then the set of all elements not belonging to $S$ is a closed set.)

Cantor also used the idea of a *limit point*, another notion that is central to topology. Given a set $S$, a point $x$ is called a limit point of $S$ whenever every open set containing $x$ also contains points of $S$ different from $x$. The point $x$ may or may not belong to $S$. The following are five examples of the idea of limit point:

Example 3.1: Dedekind's construction, which is described in the section *Richard Dedekind and the Continuum*, showed that every real number is a limit point of the set of rational numbers.

Example 3.2: The set of limit points of $\{x: 0 < x < 1\}$ is the set $\{x: 0 \leq x \leq 1\}$. (To see why, notice that any circle drawn about any point in $\{x: 0 \leq x \leq 1\}$ will contain points belonging to $\{x: 0 < x < 1\}$, so every point in $\{x: 0 \leq x \leq 1\}$ is a limit point of $\{x: 0 < x < 1\}$. On the other hand, choose a point $x$ that is not in $\{x: 0 \leq x \leq 1\}$. It is always possible to draw a circle about $x$ that does not contain any point in $\{x: 0 < x < 1\}$. Therefore, no point outside of $\{x: 0 \leq x \leq 1\}$ is a limit point of $\{x: 0 < x < 1\}$.)

Example 3.3: The set of limit points of $\{x: 0 \leq x \leq 1\}$ is $\{x: 0 \leq x \leq 1\}$. (The proof is similar to that of the preceding example.)

Example 3.4: The set of limit points of $\{1, \frac{1}{2}, \frac{1}{3}, \ldots\}$ is the set consisting of the number 0. (Any circle that extends $r$ units to the right of zero will contain every number of the form $\frac{1}{n}$, where $n$ is greater than $\frac{1}{r}$.)

Example 3.5: The set of limit points of the set of natural numbers is the empty set.

Cantor was a pioneer. He saw further than his contemporaries. Although some very prominent mathematicians of Cantor's time opposed his work because they found his ideas to be so foreign, today many of those mathematicians are remembered principally for their opposition to Cantor. Meanwhile, Cantor's insights have become central to mathematical thought in general and topology in particular.

## Different-Looking Sets, Similar Properties: Part I

When Cantor discussed sets of "points," it seems clear, at least in retrospect, that he usually meant points on a line or points in a plane or points in three-dimensional space—in other words, geometric points, but many sets, even many mathematical sets, have nothing at all to do with geometry in the usual sense. We can form sets of functions, or sets of letters, for example, and *purely as a matter of convention* we can call the elements in these sets "points." Furthermore, we can take subsets of these sets, and if (somehow) we can find a way to define what it means for a point to be an interior point of a subset, then we can talk about open sets, closed sets, limit points, and a host of other related notions. We can, in effect, divorce the notion of *point* from any geometric interpretation, and this is exactly what happened.

One interesting generalization of the concept of point can be found in the work of the Italian mathematician Giulio Ascoli (1843–96). To better appreciate Ascoli's idea, we begin with the work of the German mathematician Karl Weierstrass (1815–97). Weierstrass began his time in college as an indifferent student at the University of Bonn, where he had planned to study law. He eventually graduated from the Academy of Münster and began his working life as a schoolteacher, and it was as a schoolteacher that he worked for the next 14 years. It was also during this time that Weierstrass became obsessed with the concept of rigor in analysis. Working in isolation, he attempted to place analysis on a firm logical footing, something it had lacked since the days of Leibniz and Newton. He eventually published some of his ideas in *Journal für die reine and angewandte Mathematik (Journal for Pure and Applied Mathematics)*, an important research publication, and not long after that he was awarded an honorary doctorate by the University of Königsberg. Later, he found a position at the Royal Polytechnic School in Berlin. Weierstrass's example of a continuous nowhere differentiable function was the first such example to become widely known, although Bolzano was the first to conceive of such a function, and today Weierstrass is often called the "father of modern analysis."

The discovery of Weierstrass that is most relevant to this chapter concerns the existence of limit points for *bounded* infinite subsets of points in the plane. (A subset of the plane is bounded if one can draw a circle about the set so that all the points lie inside the circle. So, for example, the unit square is a bounded set, but the set $\{(n, 0), n = 1, 2, 3, \ldots ,\}$ is not bounded.) Today, Weierstrass's discovery is usually stated for $n$-dimensional spaces, where $n$ can represent any natural number, but in 1865, when Weierstrass originally revealed his discovery, he stated it in terms of sets of points in the plane. His theorem: "Every infinite bounded subset of the plane has a limit point." In other words, if $S$ is any subset of the plane that is bounded and contains infinitely many points, there is a point in the plane, which we will call $P$—and $P$ may or may not belong to $S$—such that $P$ is a limit point of $S$. This theorem, generalized to $n$-dimensional spaces, is now called the Bolzano-Weierstrass theorem.

Returning to Ascoli, in 1884, he published a paper in which he considered a set of functions defined on the closed interval: $\{x: a \leq x \leq b\}$. These functions satisfied the following criteria:

- They are *real-valued*. (A function is "real-valued" if its range is some subset of the real numbers.)

- They are *equicontinuous*. (Equicontinuity is a concept that applies to sets of continuous functions. A set of functions is equicontinuous if, given a positive number, which we will call $\varepsilon$, there exists another positive number, which we will call $\delta$, such that whenever the distance between $x_1$ and $x_2$ is less than $\delta$, the distance between $f(x_1)$ and $f(x_2)$ will be less than $\varepsilon$ *and the same $\delta$ and $\varepsilon$ will work for every pair of points $x_1$ and $x_2$ in the domain and for every function f in the set.*)

Let $I$ represent the set of all equicontinuous real-valued functions on $\{x: a \leq x \leq b\}$. One way of understanding Ascoli's theorem is that it takes the Bolzano-Weierstrass theorem and applies it to $I$. The set $I$ corresponds to the plane in the Bolzano-Weierstrass theorem, and the "points" of $I$ are equicontinuous functions.

To continue with the analogy between the plane and the set $I$, Ascoli needed a way to determine the distance between two points of $I$. He defined the distance between two functions in $I$ as the maximum (vertical) distance between their graphs. Having defined the distance between two functions, Ascoli could "draw a circle" about each point in $I$. Let $g$ be a point—that is, a function—that belongs to $I$. By "a circle centered at $g$," we mean a subset of $I$ of the form $\{f: -M \leq g(x) - f(x) \leq M\}$ for some positive number $M$. Each such $M$ determines a "circle" centered at $g$, where $M$ is, in effect, the radius of the circle. Let $S$ represent an infinite subset of $I$. Just as Weierstrass required that his infinite point set in the plane be bounded, Ascoli also required that his set $S$ be bounded. In Ascoli's case, that means that there is a single positive number $M$ such that $-M \leq f(x) \leq M$ for every element $f$ in $S$. (This condition describes a "circle" centered at the "point" $h(x)$ defined by the condition $h(x) = 0$ for all $x$ in the interval $a \leq x \leq b$.) Again, the idea is exactly the same as in the Bolzano-Weierstrass theorem.

Ascoli showed that whenever the infinite set $S$ is bounded, the space $I$ contains a function that is a limit point of $S$. In other words, there is a function $g$ belonging to $I$—and $g$ may or may not belong to $S$—such that every circle centered at $g$ will also contain elements of $S$ different from g. (In fact, it will contain infinitely many elements of $S$.)

To summarize: The set $I$ corresponds to the plane. Ascoli's bounded set $S$ corresponds to the bounded set $S$ in the Bolzano-Weierstrass theorem, and the functions belonging to $I$—or, more properly, the *points* of $I$—correspond to the points in the plane. Ascoli generalized the notion of point, which had previously been a geometric concept, to include functions. Ascoli's "space" is a space of functions. He showed that two very different-looking spaces of points share an important property: In both spaces, infinite bounded sets have limit points. Ascoli's discovery raised the question of what other properties might be shared by very different-looking spaces of points, and this is exactly the kind of question that topology enables one to answer.

## Different-Looking Sets, Similar Properties: Part II

Another interesting generalization of the idea of "point" was proposed by the French mathematician Émile Borel (1871–1956). Borel was an extremely productive mathematician and a staunch French nationalist. Drawn to mathematics at an early age, he began to publish mathematical papers while still an undergraduate. He received a Ph.D. at the age of 22, and for years thereafter, he averaged a mathematical paper every few months. He also wrote several books. During World War I, he volunteered to fight for France as an officer in an artillery battery. After the war, he continued his research into mathematics and also served in the French government—first in the national legislature, the Chamber of Deputies, and in the late 1930s, as minister of the Navy. He was imprisoned after the German conquest of France and later joined the French Resistance. Mathematically, Borel is best remembered for his contributions to the theory of probability, set theory, and measure theory. His name is also attached to an important topological result, which is described in chapter 5.

In this chapter, we recall two of Borel's contributions. The first contribution of interest is not widely remembered today, but it was an interesting generalization of the notion of point. This paper, which was published in 1903, considered a set the elements of which were lines in the plane. In other words, whereas Cantor had investigated sets of geometric points, Borel investigated sets in which lines took the place of geometric points. He defined a method for computing the distance between any two lines in the set. A function that is used to compute the distance between pairs of points in a set is called a *metric*. Borel used his metric to define the concept of "closeness" as it applied to his "space" of lines. With the help of his metric, Borel could investigate questions about the structure of his set. He could answer questions such as the following:

- Given a set $S$ of lines and a line $l$ belonging to $S$, is $l$ an interior point of $S$?

- Given a set $S$ of lines and an arbitrarily chosen line $l$, is $l$ a limit point of $S$?

- Given a set $S$ of lines, is $S$ open? Closed? Neither? (Unlike doors, in topology the words closed and open are not antonyms. As demonstrated in chapter 5, a set can, for example, be open and closed simultaneously.)

Borel's paper is a nice illustration of some of the ways that mathematicians of the time were generalizing the set theoretic discoveries of Cantor. In his 1903 paper, Borel also included a conceptually similar analysis of a set of planes in three-dimensional space. In this space, planes played the role of points. Again, he defined a metric on the set of planes and asked and answered a number of interesting questions about limit points, interior points, and so forth. On the one hand, it may be difficult to see immediate practical applications for Borel's ruminations about the properties of sets of lines in the plane and sets of planes in three-dimensional space, but his paper helped mathematicians further generalize the concept of point. This was an important consideration at the time, and a highly abstract conception of the term *point* is now at the center of mathematical thought.

The other contribution Borel made to the development of topology that is of interest to us concerns his very important generalization of the concepts of *open set* and *closed set*. Recall that an open set is defined as a set with the property that every element in the set is an interior point of the set. Sets, whether or not they are open, may be combined by taking their union. (The union of the two sets $A$ and $B$ is the set consisting of all the elements of $A$ and all the elements of $B$. The union of $A$ and $B$ is written as $A \cup B$.) As a matter of definition, if $P$ is an interior point of the open set $A$, it will be an interior point of any set to which $A$ belongs. In other words, if $P$ lies in the interior of $A$, it will also lie in the interior of any set that contains $A$. As a consequence, the union of any collection of open sets must be an open set.

Consider, for example, the collection of open sets $\{x: 0 < x < 1\}$, $\{x: -1 < x < 0\}$, $\{x: 1 < x < 2\}$, $\{x: -2 < x < -1\}$, . . . Each set in the collection consists of all the real numbers between two adjacent integers, and each set is an open set. Consequently, the union of all such sets is open. (Another logical consequence of this example is that the set of integers, which is the set of all numbers not belonging to the union, is a closed subset of the set of all real numbers. Why? As a matter of definition—see page 46—the set of all elements not belonging to an open set is a closed set.)

The *intersection* of any finite collection of open sets is an open set, but the intersection of an infinite collection of open sets may or may not be an open set. (The intersection of a collection of sets consists of exactly those points that belong to all of the sets in the collection.) Consider, for example, the collection of open sets $\{x: -\frac{1}{n} < x < +\frac{1}{n}\}$, where $n$ can represent any natural number. First, notice that each such set is open, but the intersection of all these sets is {0}, and the set consisting of the single number zero is a closed set. (To see why this is true, choose any number $x$ different from zero. Draw a circle with $x$ as its center and with a radius so small that the number zero lies outside the circle. This demonstrates that the set of all nonzero numbers is open because each element in the set is an interior point of the set. Consequently, the set consisting of the number zero is closed.)

Borel defined a new class of sets called "$G$-delta" sets, which are represented by the symbol $G_\delta$. A $G_\delta$ set is a generalization of the concept of an open set. A $G_\delta$ set is any set that can be represented as the intersection of a countable intersection of open sets. This includes all open sets and many sets that are not open.

In a similar way, Borel generalized the concept of closed set. Because the union of any collection of open sets is open, it follows that the intersection of any collection of closed sets is closed. To see why, let $A$ and $B$ be open subsets of the real line. Let ~$A$ represent the set of all points on the real line that do not belong to $A$. The set ~$A$ is closed because $A$ is open. (The expression ~$A$ is pronounced "the *complement* of $A$.") Let ~$B$ represent the complement of $B$. Again, because $B$ is open, ~$B$ is closed. Finally, let ~ $(A \cup B)$ represent the complement of the union of $A$ and $B$. A famous set theoretic "law" that many students encounter in their high school math classes—one of the so-called *de Morgan's laws*—states that the complement of the union of two sets equals the intersection of the complements. In symbols, this is written in the following way:

$$(\sim A) \cap (\sim B) = \sim (A \cup B)$$

Because $A$ and $B$ are open, so is $A \cup B$, and because $A \cup B$, is open, its complement is closed. Therefore, the intersection of the two closed sets ~$A$ and ~$B$ is closed. This is an example of the "set-theoretic calculus." (A very sophisticated version of the set-theoretic calculus was developed early in the 20th century by the Polish topologist Kazimierz Kuratowski.)

Just as the countable intersections of open sets may or may not be open, the countable union of closed sets may or may not be closed. Borel defined another category of sets, a generalization of the concept of closed set, called an "$F$-sigma" set, and it is denoted with the symbol $F_\sigma$. (The symbol $\sigma$ is the Greek letter sigma.) Every $F_\sigma$ can be written as the countable union of closed sets.

The collection of open sets, closed sets, $G_\delta$ sets, and $F_\sigma$ sets (as well as generalizations of the $G_\delta$ and $F_\sigma$ sets) form the collection of what are now known as Borel sets. Borel sets are used extensively in probability theory, topology, and analysis.

This was the situation in the early years of the 20th century: Mathematicians were now faced with a dizzying array of sets. There were, for example, sets of geometric points, sets of functions, sets of lines, and sets of planes, and on each of these sets they imposed structure in the form of open sets and closed sets, metrics, limit points, $G_\delta$ sets, and $F_\sigma$ sets. They discovered many interesting relations among these sets. Their research revealed properties that some sets of points have but that other sets of points do not have. They gave names to these newly discovered properties, and they investigated the logical implications of their discoveries.

At a very basic level, early topologists were seeking criteria that would enable them to determine when two different-looking sets are fundamentally the same and when they are different. This is just what Euclid had done more than 2 millennia earlier. Euclid's answer to the problem of applying the concept of "sameness" to a collection of figures was what we now know as the set of Euclidean transformations. If, using any series of Euclidean transformations, one object can be made to coincide with another, then they are the same; otherwise, they are different. Topologists sought a conceptually similar criterion for determining when sets are the same, but because the sets imagined by 19th- and early 20th-century mathematicians were often so different looking—sets of geometric points versus sets of lines in the plane versus sets of functions, for example—they needed to identify a set of transformations that were quite different from those of Euclid. These are now called topological transformations. These mathematicians also sought to identify exactly those properties that are preserved under the set of topological transformations, and, out of this effort, they created the discipline of set-theoretic topology.

# 4

# THE FIRST
# TOPOLOGICAL SPACES

Mathematical knowledge agglomerates. When progress is made, mathematical discoveries are published for all to see, and the new is added to the old. In mathematics, new knowledge does not replace old knowledge, it grows alongside it. This is different from the way science often progresses. In science, new discoveries often replace old ones. The Renaissance-era discovery that the Earth orbits the Sun replaced the more ancient belief that the Sun orbits the Earth. The ancient Earth-centered model was abandoned because it was incorrect. By contrast, mathematical knowledge is never abandoned. No mathematical discovery can show Euclidean geometry is incorrect, because Euclidean geometry is logical, and the sole criterion for truth in mathematics is logic. A statement that is a logical consequence of a set of axioms remains a logical consequence for all time. As new mathematical discoveries are added to the old, they create a more diverse store of mathematical knowledge.

By 1910, the store of mathematical knowledge had become very diverse, indeed. In particular, Cantor's theory of sets had become an important and independent branch of mathematics, but its importance extended far beyond questions about the nature of sets of geometric points. Researchers in many different fields began to express their insights using the ideas and vocabulary created by Cantor—cardinality, one-to-one correspondences, open sets, closed sets, interior points, limit points. His insights had provided a basis for a new and unifying conception of mathematics. Sets were not just passive collections of objects; they had structure, and researchers began to use those structures to compare and contrast sets. They wanted to understand the relationships that exist among very different-looking sets of points, and with their more abstract conception of the word *point*, almost everything could be described as a set of points. They wanted to know when different-looking sets were fundamentally the same and when they were fundamentally different. Faced with diversity, they sought unity. As they attempted to understand sets at their most basic level, they created the field of set-theoretic topology as it is understood today.

## Felix Hausdorff and the First Abstract Topological Spaces

The German mathematician Felix Hausdorff (1868–1942) was a pioneer in the attempt to create useful abstract topological spaces. He was born in Breslau in the former state of Prussia—now Wrocław, Poland—and grew up in Leipzig, Germany. He studied mathematics and astronomy at the University of Leipzig, and his Ph.D. thesis was about a topic in astronomy. In addition to his work in topology, he also wrote a number of literary works. Under the pseudonym Dr. Paul Mongré, Hausdorff wrote poetry, philosophy, and even a play. Philosophically, he was an admirer of the German philosopher and critic Friedrich Nietzsche, and in his writings about philosophy, Hausdorff speculated extensively about the nature of time. His play was a comedy about two men and their plans to fight a duel, and it was a success. At a theater in Berlin, the play had a run of more than 100 performances. Historically speaking, not many mathematicians have written successful comedies.

Today, Hausdorff is remembered primarily for his work in topology and for the tragic way his life ended. Despite international recognition for the quality of his work in topology, Hausdorff, who was Jewish, lost his job in 1935 as a result of the policies of the National Socialist Party (Nazis). Unlike many of his Jewish contemporaries in academia, Hausdorff, who certainly could have left Germany, remained at home. When no German journal would publish his papers, he published them in *Fundamenta Mathematicae*, a Polish mathematical journal that emphasized topological research. In 1942, when the authorities ordered Hausdorff, his wife, Charlotte Hausdorff, and his sister-in-law, Edith Pappenheim, to move to a concentration camp, the three of them committed suicide. They are buried in Bonn, Germany.

In topology, Hausdorff's most influential work, which is almost always identified by its German name, is entitled *Grundzüge der Mengenlehre* (Basics of Set Theory). Hausdorff published his treatise in 1914, and it contains the first detailed axiomatic investigation of abstract topological spaces. It begins by defining a topological space as a set $X$ of "points" together with a collection of subsets of $X$ called *neighborhoods*. (Today, the term *neighborhood* is often used synonymously with the term *open set*.) In Hausdorff's model, the collection of neighborhoods satisfies the following four axioms:

*Felix Hausdorff was one of the first to create "abstract" topological spaces. His work permanently changed the history of the subject.* (University of Bonn)

Axiom 1. For each $x$ in $X$, there is at least one neighborhood, which we will call $U$, containing $x$.

Axiom 2. Let $U$ and $V$ be two neighborhoods of the point $x$. The intersection of $U$ and $V$ contains a neighborhood of $x$. (The intersection of the two neighborhoods, which is written symbolically as $U \cap V$, is the set of points belonging to both $U$ and $V$.)

Axiom 3. Suppose that $U$, which is a neighborhood of $x$, also contains the point $y$, then the neighborhood $U$ also contains a neighborhood of $y$.

Axiom 4. If $x$ and $y$ are distinct points in $X$, then there exist neighborhoods $U$ of $x$ and $V$ of $y$ such that the two neighborhoods share no common points. In symbols, this idea can be expressed in the following way: $U \cap V = \varnothing$ for some choice of neighborhoods $U$ and $V$.

Hausdorff's conception of a topological space is not quite the same as the one currently in use, but his axiomatic development of the subject is conceptually similar to the way topology is expressed today. It is also conceptually similar to the way Euclid developed geometry in *Elements*. To better understand topology and its history, it is worthwhile to develop an appreciation for what Hausdorff's axioms actually imply. Later in the volume, we will compare his axioms to the ones in general use today. First, consider each of the following examples:

Example 4.1: Let $X$ represent the real line. To make $X$ a topological space (as Hausdorff understood it) we need to identify the neighborhoods. This is easier if we introduce some notation. Let $U(x, r)$ stand for the set of real numbers within $r$ units of the point $x$. The letter $x$ represents an arbitrary real number, and the letter $r$ represents an arbitrary positive real number. In symbols, we can write $U(x, r) = \{t: x - r < t < x + r\}$. Or to put it still another way, if we draw a small circle about $x$ of radius $r$, $U(x, r)$ contains all the points on the real line that lie within this circle. Every positive number $r$ gives us another neighborhood of $x$. Now let us check that the set $X$ together with the set of all such neighborhoods $U(x, r)$ satisfy each of Hausdorff's axioms:

1. Every point $x$ of the real line belongs to some—in fact, infinitely many—neighborhoods of the form $U(x, r)$, so Hausdorff's first axiom is satisfied.
2. To see that Hausdorff's second axiom is satisfied, let $U(x, r)$ and $U(x, s)$ be two neighborhoods of the point $x$. Their intersection will also contain a neighborhood of $x$. (Because both intervals are centered at $x$, the smaller of the two intervals will satisfy the requirement.)

3. Consider the neighborhood $U(x, r)$, and let $y$ be a point belonging to this neighborhood, then there is a neighborhood of $y$ that also lies in $x$. To see this, suppose that $y$ lies to the right of $x$ so that $x < y < x + r$. Let $s$ be a positive number so small that $y + s < x + r$, then the neighborhood $U(y, s)$ lies within $U(x, r)$. (If $y$ lies to the left of $x$, then choose $s$ to be so small that $x - r < y - s$. For this value of $s$, $U(y, s)$ will belong to $U(x, r)$.) See the accompanying illustration.

4. If $x$ and $y$ are any two distinct real numbers, we can always draw two small circles, one centered at $x$ and one centered at $y$, that do not overlap. Choose one neighborhood of $x$ that lies entirely within the circle about $x$, and choose one neighborhood of $y$ that lies entirely within the circle about $y$. These neighborhoods satisfy Hausdorff's fourth axiom.

Example 4.2: Let $X$ be the interval $\{x: 0 < x < 1\}$. Use the same neighborhoods that were defined in Example 4.1, but discard any neighborhood from Example 4.1 that extends beyond the endpoints of $\{x: 0 < x < 1\}$. Hausdorff's axioms are satisfied, and the proof is identical to that found in Example 4.1.

Example 4.3: Let $X$ be the union of these two intervals: $\{x: 0 < x < 1\}$ and $\{x: 2 < x < 3\}$. Use the same neighborhoods that were defined in Example 4.1, but discard any neighborhoods from Example 4.1 that contain points outside the two intervals. Hausdorff's axioms are satisfied, and the proof is identical to that found in Example 4.1.

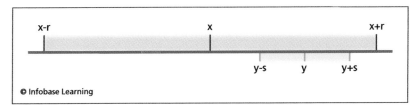

© Infobase Learning

*The set of all real numbers together with the collection of subsets of the form $\{t: x - r < t < x + r\}$ for every positive real number $x$ satisfies Hausdorff's axioms. This picture illustrates that axiom 3 is satisfied.*

Example 4.4: Let $X$ denote the plane. Define a neighborhood of the point $(x, y)$ to be the interior of a circle of radius $r$ centered at $(x, y)$. With a slight modification of the notation in Example 4.1, we can write $U((x, y), r)$ to represent these sets. The set of all such neighborhoods for every point $(x, y)$ and for every positive $r$ will satisfy Hausdorff's axioms. (This can be verified as in Example 4.1. All that is necessary is to use the distance formula for the plane in place of the distance formula for the line and replace the word *interval* with the word *disc*.)

Example 4.5: Let $X$ denote that subset of the plane consisting of all points less than 1 unit from the origin. (This set is a disc of radius 1.) Define the neighborhoods as in Example 4.4, but discard any neighborhoods from Example 4.4 that contain points outside $X$. Hausdorff's axioms are satisfied.

Example 4.6: Let $X$ represent three-dimensional *Euclidean space*. A neighborhood of a point $(x, y, z)$ in $X$ will be the interior of a sphere of radius $r$, where $r$ is any positive number. We can even modify the notation of Example 4.1 again and write $U((x, y, z), r)$ for each such neighborhood. The set of all such neighborhoods satisfies Hausdorff's axioms. (The proof is [again] just as in Example 4.1—just use the distance formula for three-dimensional space instead of the distance formula for the real line, and use the phrase *interior of a sphere* instead of the word *interval*.)

Example 4.7: The open ball of radius 1 centered at the origin is defined as the set of all points in three-dimensional space that are less than 1 unit from the origin. Let the open ball of radius 1 be $X$. Use the neighborhoods as defined in Example 4.6, but discard any neighborhoods that extend beyond the surface of the ball. Hausdorff's axioms are satisfied.

Example 4.8: The set of three points shown in the accompanying diagram satisfies Hausdorff's axioms when the neighborhoods are defined by the indicated curves.

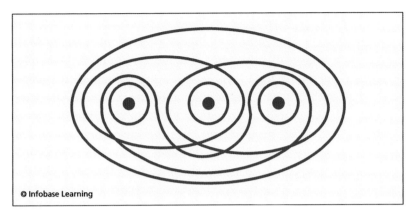

*This set of three points satisfies Hausdorff's axioms when the neighborhoods are defined by the curves.*

Having stated his axioms for a topological space, there were, logically speaking, several directions in which Hausdorff could go. First, he investigated the logical consequences of his axioms. All spaces that satisfy the axioms also share all properties that are logical consequences of the axioms. As the preceding examples illustrate, this is a very large class of different-looking spaces, and Hausdorff did draw a number of deductions from his axioms. In some ways, however, this approach is too broad. It makes no distinctions among spaces that are in some ways very different from one another. The identification of broad patterns is an important part of mathematical research, but details also matter. In addition, it is important to investigate the characteristic properties of more narrowly defined classes of spaces, and this Hausdorff also did.

After Hausdorff investigated some of the logical implications of his original set of four axioms, he added additional axioms to the original four. This had the effect of restricting the class of spaces to which his results applied, but it allowed him to study the characteristic properties of spaces that are of special interest. By way of example, he added an axiom that stated that around each point $x$ in the set $X$ there exists, at most, a countable collection of neighborhoods. (None of the topologies described in Examples 4.1 through 4.7 satisfy this axiom, because the set of neighborhoods assigned to each $x$ in $X$ can be placed in one-to-one correspondence with the uncountable set of real numbers, but we could modify the topologies so they conform to this requirement. We could, for example, restrict $r$ to lie within the set $\{1, \frac{1}{2}, \frac{1}{3}, \ldots\}$. It is a simple matter to verify that, with this new restriction, all of the spaces in Examples 4.1 through 4.7 would again satisfy Hausdorff's original axioms as well as the added requirement that the neighborhoods about each point form a countable set. We can obtain the verification by repeating everything word-for-word and keeping in mind the new restriction on $r$. Of course, Example 4.8 satisfies Hausdorff's "countability axiom" because the number of neighborhoods about each point is finite.)

At another point in his treatise, Hausdorff introduced an axiom that required that the set of all neighborhoods about all the points form a countable set. The topologies of Examples 4.1 through 4.7 again fail to satisfy this axiom, because the cardinality of the set of neighborhoods in these examples must be at least as large as the cardinality of $X$, and as Cantor demonstrated (see chapter 3), the cardinality of the set of points on an interval, on the line, in the plane, in a disc, in three-dimensional space, or in a (three-dimensional) ball is larger than the cardinality of the set of natural numbers.

Hausdorff supplemented the axioms in other ways as well, and this allowed him to further narrow the class of spaces to which his results applied. Using his new topological methods, Hausdorff was able to provide a new proof of the Bolzano-Weierstrass theorem (see chapter 3, pages 48 and 49) as well as proofs of a number of other important theorems. Hausdorff's emphasis on a careful axiomatic development is completely classical. His approach to mathematical research would have been familiar to Euclid, although the subject matter would probably have seemed strange to the ancient geometer.

Hausdorff also established a criterion for determining when two different-looking collections of neighborhoods defined on the same set $X$ determine the same topological characteristics. When, for example, do two different-looking collections of neighborhoods determine the same open sets? Not every different-looking definition produces a different collection of open sets. To appreciate the problem of topological equivalence, recall that in Example 4.4, the neighborhoods of each point $(x, y)$ in the plane were defined as the interiors of circles centered at $(x, y)$. (We will call this the "circle topology.") Suppose, instead, that we place a Cartesian coordinate system on the plane and define the neighborhoods for each point $(x, y)$ to be the interiors of squares centered at $(x, y)$, with the additional property that the sides of the squares are parallel to the coordinate axes. (We will call this the "square topology." The plane with this set of neighborhoods also satisfies Hausdorff's axioms.) Do the two different-looking ways of defining neighborhoods produce the same open sets, the same limit points, and so on?

To be specific, let $S$ be an open subset of the plane with respect to the circle topology—in other words, given any point $x$ in $S$, there is a small disc centered at $x$ and lying entirely within $S$. Is $S$ also an open subset of the plane in the square topology? In other words, given any point in $S$, is there also a small square centered at the point $x$ that lies entirely within $S$? The answer is yes, because inside any circle we can always draw a square so if the circle lies within $S$ so does the square. Similarly, suppose $x$ is an interior point of the set $S$ in the square topology. This means that we can draw a square about $x$ so that the interior of the square lies entirely within $S$. The point $x$ will also be an interior point of $S$ in the circle topology because inside any square we can draw a circle. Therefore, a point is an interior point of $S$ in the circle topology if and only if it is an interior point of $S$ in the square topology. This illustrates the fact that different-looking neighborhoods can produce identical topological characteristics.

More formally, Hausdorff developed the following equivalence criterion: Given one set $X$ with two different collections of neighborhoods, which we will call $\{U_i\}$ and $\{V_j\}$ (the subscripts are just there to indicate that there is more than one neighborhood in each collection), the two collections of neighborhoods are equivalent if:

1. for every point $x$ and every neighborhood $U_i$ of $x$, there is a neighborhood $V_j$ of $x$ contained in $U_i$, and

2. for every point $x$ and any $V_j$ containing $x$, there exists a $U_k$ contained in $V_j$.

The details of the shapes of the neighborhoods are not important. If the two sets of neighborhoods satisfy the equivalence criterion, then any set that is open, closed, and so on with respect to one set of neighborhoods will be open, closed, and so on with respect to the other set of neighborhoods.

Hausdorff also generalized Bolzano's definition of continuous function. Hausdorff's definition is expressed solely in terms of neighborhoods. This is important, because neighborhoods can be defined without reference to the distance between points. In fact, for some sets, it is impossible to define the concept of distance. Here is how Hausdorff understood the concept of continuity: Let $X$ and $Y$ be two topological spaces, which in this case means that they are sets of points together with collections of neighborhoods satisfying Hausdorff's axioms. Let $f$ be a function with domain $X$ and range in $Y$. Let $x_1$ be a point in $X$. The function $f$ is continuous at $x_1$ if for any neighborhood $V$ containing $f(x_1)$, there is a neighborhood $U$ containing $x_1$ such that $f(x_2)$ belongs to $V$ whenever $x_2$ belongs to $U$. Another way of saying the same thing is that the function $f$ *transforms* $U$ into $V$. Hausdorff's definition of continuity can be stated in terms of a transformation: Suppose we are given a neighborhood $V$ of $f(x_1)$. If it is always possible to find a neighborhood $U$ of $x_1$ such that $f$ transforms $U$ into $V$, then $f$ is continuous at $x_1$. Hausdorff expressed the concept of "closeness" in terms of neighborhoods rather than distances, but when neighborhoods are expressed in terms of distances, Hausdorff's and Bolzano's ideas are completely equivalent.

To illustrate how Hausdorff's definition of continuity is equivalent to Bolzano's definition when they both apply, suppose that $f$ is a real-valued function with a domain that is a subset of the real numbers. If $f$ is continuous according to Bolzano's definition, then for each $x_1$ in the domain and for each positive number $\varepsilon$, there exists a positive number $\delta$ such that whenever the distance from $x_2$ to $x_1$ is less than $\delta$, then the distance from $f(x_1)$ to $f(x_2)$ will be less than $\varepsilon$. In other words, suppose we are given a neighborhood—that is, an open interval—of width $2\varepsilon$ centered about $f(x_1)$. Call this interval $V(f(x_1), \varepsilon)$. If there is a neighborhood centered about $x_1$ of width $2\delta$—call this neighborhood $U(x_1, \delta)$—such that

$f$ transforms $U(x_1, \delta)$ into $V(f(x_1), \varepsilon)$, then $f$ is continuous at the point $x_1$. This is just Hausdorff's definition of continuity. If, on the other hand, we start with Hausdorff's definition of continuity and we assume $f$ is continuous according to Hausdorff, we can prove that $f$ must also be continuous according to Bolzano's definition of continuity. (The proof is obtained by reading this paragraph, more or less, backward!) The advantage of Hausdorff's definition is that it works for a larger class of spaces than does Bolzano's.

It is worth noting that there are no computations in Hausdorff's work. This is characteristic of topology, which is concerned with the most basic properties of sets. The sets may consist of numbers, or geometric points, or functions, or dots with curves drawn around them, as in Example 4.8 of this section. Although topology grew out of geometry—at least in the sense that it was initially concerned with sets of geometric points—it quickly evolved to include the study of sets for which no geometric representation is possible. This does not mean that topological results do not apply to geometric objects. They do. Instead, it means that topological results apply to a very wide class of mathematical objects, only some of which have a geometric interpretation.

Finally, it is important to point out that not all the ideas described in this section originated with Felix Hausdorff. As with most acts of discovery, there were others who prepared the way, but Hausdorff was a pioneer in the subject. He examined many mathematical "objects," such as the real number line, various subsets of the line, the plane, various subsets of the plane, and so on, and from these many examples he created abstract models of these not-quite-as-abstract spaces. The models were created to retain those properties of the spaces that were important to Hausdorff. This sort of mathematical modeling is another way of understanding what it is that mathematicians do when they do mathematics.

Much of mathematics can be understood as "model building." In some cases, this is obvious. Applied mathematicians, for example, routinely describe their work as *mathematical modeling*. They "model" the flow of air over a wing, the impact of a country's national debt on the future growth of its economy, the spread of disease throughout a population, and so forth. They make certain assumptions about the system in which they are interested, and then they investigate the logical consequences of their assumptions. They are successful when their models are simple enough to solve but sophisticated enough to capture the essential characteristics of the systems in which they are interested.

Some so-called pure mathematicians also build models. They examine many different mathematical systems and attempt to identify those properties that are important to their research. They isolate those properties by specifying them in a set of axioms, and then they investigate the logical consequences of the axioms. This is what Hausdorff did. The difference between what pure mathematicians do and what applied mathematicians do is apparent only at the end of the process. The work of applied mathematicians must pass one more test before it can be judged successful. It must agree with observations and experiments. If their work fails to conform to experimental results and observations, it must be rejected because it is not useful. It did not fulfill its intended purpose. It must be rejected even when it is logically coherent. By contrast, the work of pure mathematicians is subject only to the rules of logic. Utility is not a concern. Any model that contains no logical errors is a successful model.

## Topological Transformations

So what is topology? It might seem that Hausdorff, who contributed so much to the foundations of the subject, would have had a ready answer. He could certainly have pointed to many examples of topological spaces. He had studied some; he had invented others, but when he published his masterpiece, *Grundzüge der Mengenlehre*, he probably would have had a hard time defining the subject that he had done so much to help to create. It took decades of work to define precisely what topologists study when they study topology, and this work was not completed when *Grundzüge der Mengenlehre* was published. To say that topologists study "topological properties" says little because it only shifts our attention away from the term *topology* and to the term *topological property*.

*From a topological viewpoint, all of these shapes are the same.* (Playthings, Inc.)

One of the earliest ideas about what it means for a property to be topological was proposed by the German astronomer and mathematician August Ferdinand Möbius (1790–1868). Möbius made a number of important contributions to astronomy and geometry. In particular, he did pioneering research on the properties of one-sided surfaces, the most famous example of which is the Möbius strip, which many students make at least once during their time in school. Möbius believed that topological properties are exactly the properties that are preserved when the surface on which they are defined is stretched, or compressed, or otherwise continuously distorted. In 1863, for example, he wrote,

> If, for example, we imagine the surface of a sphere as perfectly flexible and elastic, then all the possible forms which one can give to it by bending and stretching (without tearing) are elementarily related to each other.

This characterization of topology is still in use today when, in the popular press, topology is described as "rubber sheet geometry."

Möbius was far ahead of his contemporaries when, already in his 70s, he proposed that mathematicians interested in *analysis situs* confine their attention to those properties that are preserved when surfaces are continuously deformed, but there are problems with this conception of topology. These problems became apparent only later as mathematicians became aware of sets more general than surfaces and transformations more general than bending and stretching. This is one reason that Peano's space-filling curve was so important to the history of the subject. Peano's function is a continuous deformation of the unit interval. He used his function to "bend and stretch (without tearing)" the unit interval until it covered the unit square. Also, consider the constant function $f(x) = 1$ for all $x$ in the unit interval. (In other words, for every $x$ in the unit interval, $f(x)$ has the value 1.) This function is also continuous. One can argue, therefore, that it "bends and stretches (without tearing)" the unit interval into a single point. Now recall Euclid's first axiom, "Things which are equal to the same thing are also equal to one another." Peano's function transforms the unit interval into the unit square. Using Möbius's idea of continuous deformation, we should conclude that the unit interval and the unit square are "the same." Similarly, we should conclude that the unit interval and the single point 1 are the same. Finally, with the help of Euclid's first axiom (if we decide to accept the axiom!), we should conclude that a point is the same as the unit square.

Later, in chapter 5, we will see that there is a sense in which the unit square, the unit interval, and the single point are the same. They are all examples of *compact sets*. Möbius's idea is not "wrong." Mathematicians could build a theory of sets in which two sets are the same if one can be transformed into the other via a continuous function. Because while the preceding examples may make it seem that any two sets would be the same in such a theory—they show, after all, that a point and a square are the same—many sets are still fundamentally different from one another, even under Möbius's definition. No continuous function exists, for example, that can transform the set $\{x: 0 \le x \le 1\}$ into the set $\{x: 0 < x < 1\}$, even though only two points separate the two sets. This is not merely to say that a continuous function has not yet been found to effect such a transformation. Rather, no such function can be found because the transformation is impossible. (See the section "Topological Property 1: Compactness" in chapter 5 for the reason why this is true.)

In any case, even in the early history of topology, most mathematicians would have agreed that a theory that asserts that a point and a square are the same is too broad a theory to be useful. They required more from a topological transformation than that it be continuous. They sought a stronger definition of "sameness," and they found it in the concept of *homeomorphism*, which is the name now used to denote a topological transformation. A topological property is, therefore, any property that is preserved under the set of all homeomorphisms. (Again, compare this to the section "Euclidean Transformations" in chapter 1, wherein a geometric property in Euclidean geometry is any property that is preserved by the set of Euclidean transformations.)

The term *homeomorphism* was introduced by the French mathematician, scientist, and philosopher Henri Poincaré (1854–1912), who was one of the most influential mathematicians of his era. Poincaré received much of his early education at home from his mother, and he demonstrated a tremendous gift for mathematics at a young age. He contributed to many branches of mathematics and physics. Algebra, analysis, and topology were some of the fields in which Poincaré did research, and while his work in physics is not as well remembered today, he developed much of the special theory of relativity independently of Albert Einstein. Poincaré's treatment of relativity was much more mathematical—which is to say that it was not as accessible as Einstein's development of the theory—but in retrospect, it is apparent that because of Poincaré's work, the special theory of relativity would have become part of the cultural fabric in the early years of the 20th century with or without Einstein. Poincaré also wrote articles and books about philosophy, science, and mathematics for a general audience, and his work was well received. (See the Further Reading section for references to two of his works written for a nonspecialist readership.)

Poincaré's concept of homeomorphism is different from the modern concept. He did not, for example, consider abstract spaces, preferring to confine his attention to what are called

$n$-dimensional Euclidean spaces, such as the real line, the plane, and higher-dimensional generalizations of these spaces. For example, the three-dimensional space that students first encounter in analytic geometry—the space in which the points are placed in one-to-one correspondence with ordered triplets of real numbers $(x, y, z)$—is an example of such a space. Poincaré defined a homeomorphism to be a continuous one-to-one transformation between ordered sets of $n$-tuples. He also required that the transformation be differentiable. His definition worked well for the applications he had in mind, but it is too restrictive for use in abstract spaces, where it may not be possible to define the concept of derivative or even the concept of coordinates.

The definition of a topological transformation (homeomorphism) in use today cannot be traced to any single mathematician. It arose as topologists became more and more abstract in their thinking. No longer were they content to phrase their thoughts in terms of curves, surfaces, and volumes. Instead, they expressed themselves in terms of abstract topological spaces, and they needed an abstract criterion to determine when these spaces were "the same." (They preferred to use the word *equivalent*.) Here is how the German mathematician Adolf Hurwitz (1859–1919) described the goal of topology in 1897: ". . . to determine all the kinds of equivalent point-sets, to partition them into [equivalence] classes, and to see the invariants of these classes." (An *invariant* is a property that is shared by all the sets in the class.) The problem, then, was to decide upon the best way to determine when two sets are equivalent.

Peano's space-filling curve, which transforms the unit interval onto the unit square, is continuous, but it is not one-to-one. On the other hand, Cantor's one-to-one correspondence between the points of the unit interval and the points of the unit square is discontinuous. An "obvious" next step was to combine the properties of Cantor's one-to-one correspondence and Peano's continuous function and investigate the use of functions that are continuous *and* one-to-one. To appreciate what this means, recall from algebra class that a *one-to-one function* is a set of ordered pairs in which every element in the domain appears exactly once, and every element in the range appears exactly once. (It is just another way of describing a one-to-one correspondence.) Mathematicians began to investigate whether continuous one-to-one functions provide a good test for equivalence. If one topological space could be transformed into another by a continuous one-to-one function, was it reasonable, these mathematicians asked, to say that the two topological spaces are equivalent?

This is a question that can be answered only by a kind of exploration. Mathematicians considered different pairs of topological spaces and different continuous one-to-one functions between the pairs of spaces, and then they debated among themselves whether their definition of equivalence was reasonable. Did the class of continuous one-to-one functions preserve those properties that the mathematicians thought were important topological properties? In many cases, continuous one-to-one functions do preserve topological properties, and it took some time and some additional "digging" to find examples for which this definition failed. Here, in general terms, is the difficulty: Suppose that we have two topological spaces, which we call $A$ and $B$, and suppose we have a function $f$ that is a continuous one-to-one function with domain equal to $A$ and range equal to $B$. At one time, many topologists would have claimed that this was sufficient to conclude that $A$ is equivalent to or "essentially the same" as $B$. But because $f$ is a one-to-one function, the *inverse function*, which is written $f^{-1}$, also exists. Recall that the domain of $f^{-1}$ is the range of $f$, and the range of $f^{-1}$ is the domain of $f$. The function $f^{-1}$ "undoes" the work of the function $f$ in the sense that if $y = f(x)$ then $x = f^{-1}(y)$, or, to put it another way, $x = f^{-1}(f(x))$. By examining different examples of continuous one-to-one functions defined on different pairs of topological spaces, mathematicians eventually uncovered pairs of topological spaces $A$ and $B$ with the following properties:

1. There exists a one-to-one continuous function $f$ transforming $A$ onto $B$, which, according to the definition they were investigating, showed that $A$ is equivalent to $B$, but

2. the function $f^{-1}$ was not continuous, and, in fact, $B$ was not equivalent to $A$ (under the old definition).

The consensus was that this situation was unsatisfactory. To go back to Euclid's first axiom, "Things which coincide with one another are equal to one another." These early topologists had uncovered pairs of topological spaces, A and B, where (under the definition of equivalence then in use) A "coincided" with B, but B did not coincide with A. Under such circumstances, they did not want to conclude that A and B are equivalent.

It is important to emphasize that mathematicians are always free to disregard any of Euclid's axioms. Euclid's *Elements* is hardly the last word on mathematics, and Euclid considered his axioms, in contrast to his postulates, to be only "common sense" notions. Presumably, what is common sense in one culture need not be common sense in another. Nevertheless, these early topologists decided that before one could say that A and B are equivalent, A had to coincide with B, *and* B had to coincide with A.

There is another possibility. Suppose that one could find two functions, each of which is continuous and one-to-one. The first function, which we will call *f*, would have domain equal to A and range equal to B. The second function, which we will call *g*, would have domain equal to B and range equal to A. By the old definition, the existence of *f* would demonstrate that A is equivalent to B, and the existence of *g* would demonstrate that B is equivalent to A. Under these circumstances, must A and B have the same topological properties? It was not until the 1920s that the Polish topologist Kazimierz Kuratowski (1896–1980) produced an example of a pair of topological spaces and a pair of one-to-one continuous functions that satisfied all of the conditions just described and with the additional property that A and B were topologically different.

In the end, mathematicians decided that the solution to the problem of determining topological equivalence was to require still more of the function *f*. Topologists eventually came to the following consensus: Two topological spaces A and B are equivalent if (and only if) there exists a one-to-one continuous function *f* with domain equal to A and with range equal to B with the additional property that $f^{-1}$ is also continuous. Such a function is now called a homeomorphism, and today topology is defined as the study of those properties of a space that are preserved by homeomorphisms.

Another way to think of the set of all homeomorphisms is as a tool. Homeomorphisms enable topologists to classify topological spaces. Two topological spaces may "look" very different. They may be defined in different ways. We may visualize the points of which the spaces are comprised in very different ways, but provided a homeomorphism exists that transforms one space into the other, the two spaces are indistinguishable from a topological viewpoint. Homeomorphisms are, then, tools for simplification. The class of all topological spaces can be organized into equivalence classes in the same way that biologists use the concept of species to classify living organisms. Suppose that a biologist knows that two organisms, which we will call A and B, belong to the same species, and suppose that the biologist can examine only A. Knowledge that B belongs to the same species as A allows the biologist to draw many accurate conclusions about the characteristics of B without seeing B. Of course, this kind of knowledge is not enough to enable the biologist to predict every physical characteristic of B with precision, but many accurate statements can be made because members of the same species share many physical characteristics.

Similarly, the knowledge that two topological spaces, which we will again call A and B, are equivalent (or homeomorphic) enables the topologist to predict many of the properties of B from an examination of A, but B need share only those properties that are preserved by homeomorphisms. Any other conclusions that one makes about B from an examination of A are completely unwarranted. For example, if A is the interval $\{x: 0 < x < 1\}$ and we know that B is homeomorphic to A, we can conclude, for example, that B is one-dimensional, but we cannot conclude that B is an interval of finite length, because lengths are not preserved under the set of all homeomorphisms.

Example 4.9. Let A denote the set $\{x: 0 < x < 1\}$, and let B denote the real line. The function $f(x) = \dfrac{(x - \frac{1}{2})}{(x - x^2)}$ is a homeomorphism from A to B. From the point of view of topology, therefore, the real line and the set $\{x: 0 < x < 1\}$ are topologically equivalent.

# The Role of Examples and Counterexamples in Topology

Set-theoretic topology is unusual among the major branches of mathematics in its heavy reliance on examples and counterexamples. When mathematicians began to consider the topological structures of abstract sets, they created a mathematical landscape that is, in many ways, bizarre. The long search for a definition of homeomorphism described in the preceding section demonstrates how topologists' intuitions often led them astray. In fact, for decades after Cantor began to investigate sets with strange and surprising properties, topologists continued to produce examples of curves, surfaces, and other objects with properties that violated almost everyone's common sense notions of what is (or, perhaps, what should be) possible. (See, for example, the sidebar *Counterexample 4: Sierpiński's Gasket* in this chapter.)

To appreciate what makes topology different, compare it with number theory, the branch of mathematics concerned with the properties of the set of integers. The history of the theory of numbers goes back to the time of the French mathematician, linguist, and lawyer Pierre de Fermat (1601–65). Mathematicians have been studying number theory ever since Fermat's time, and the subject continues to attract attention today. Some problems have even become somewhat famous. Fermat's Last Theorem, for example, which states that there are no natural number solutions to the equation $x^n + y^n = z^n$ when $n$ is a natural number greater than 2, is one of the best-known number theoretic problems. (When $n = 2$, for example, many solutions exist. Here is one: $x = 3$, $y = 4$, and $z = 5$.) Fermat first described the problem, and he famously wrote in the margin of a book that he had found a proof that no natural number solutions existed for $n$ greater than 2 but that the margin of the book was too narrow for him to write the proof. For the next several centuries, mathematicians attempted to prove Fermat's statement. Late in the 20th century, a proof was finally published, but few people were surprised that the result was true. Perhaps more surprising is that the proof required about 200 pages of terse mathematics to establish its truth. Still, even if Fermat had gotten it wrong—even if it had turned out that there are three natural numbers $a$, $b$, $c$, and $n > 2$ such that $a^n + b^n = c^n$, it is difficult to see how this would have violated anyone's common sense notions of what is possible in mathematics.

Another example of a famous problem from number theory is Goldberg's conjecture, which states that every even integer greater than 2 can be written as a sum of two prime numbers. (A prime number is a natural number that is only evenly divisible by itself and 1. By way of example: $12 = 5 + 7$.) The problem dates to the 18th century, and although many have tried, no one has been able to prove the conjecture true or false. If Goldberg's conjecture is ever resolved—no matter whether the result is true or false—it is (again) hard to imagine the result violating anyone's common sense notions of what is possible in mathematics. As with so many problems in number theory, Goldberg's conjecture is easy to state, easy to understand, and difficult to solve, but neither the problem nor the solution (if one is ever found) is likely to be especially surprising.

By contrast, during the last years of the 19th century and for the first few decades of the 20th century, topologists regularly produced examples of functions and of sets that previous generations of mathematicians would have described as impossible. In fact, many topologists spent much of their professional careers producing examples of such spaces and functions. Recall that Bolzano and Weierstrass produced examples of functions that were continuous everywhere but nowhere differentiable. Their functions demonstrated that continuity is a concept quite distinct from differentiability, something that would surely have surprised Leibniz and Newton. (Newton, especially, would have been surprised because he visualized derivatives as velocities, and a point that moved along Bolzano's curve would have had position but no *velocity*—not, at least, in the sense that Newton understood the term.) Or consider the problem of curves. Our experience tells us that curves are fairly simple objects in the sense that they are "linelike." Branch points, which, loosely speaking, can be defined as points where the curve "splits"—one line in and two or more lines out—are exceptional, but the Polish mathematician Wacław

*(text continues on page 79)*

# COUNTEREXAMPLE 4: SIERPIŃSKI'S GASKET

What are the properties that all planar curves share? Identifying exactly what it is that makes a curve curvelike is no easy task. One early definition of a curve, described a curve as the graph of a continuous function with domain equal to the unit interval. Peano, however, showed that under this definition the unit square, including all points in its interior, is a curve. Other definitions were proposed and eventually rejected. In 1915, the Polish mathematician Wacław Sierpiński (1882–1969) explored the prevailing definitions of curves by producing a curve with a very strange property, indeed. His curve is called the Sierpiński gasket.

To appreciate what is peculiar about the Sierpiński gasket, we begin by considering an arbitrary planar curve. Let $S$ denote the set of points that constitute the curve. We can partition $S$ into three disjoint subsets, which we will call $S_1$, $S_2$, and $S_3$. (The word *partition* means that every element of $S$ will belong to one of these three sets, and no element will belong to more than one of these sets.) The set $S_1$ contains the endpoints of $S$. A curve may have several endpoints, or it may have none. (An asterisklike object, for example, is a curve with several endpoints, and a circle is a curve with no endpoints.) To test whether a point $x$ belonging to $S$ is an endpoint of $S$, imagine drawing small circles, each of which is centered at $x$. If each sufficiently small circle intersects $S$ at exactly one point, then $x$ is an endpoint, and it can be assigned to $S_1$.

A point $x$ in $S$ belongs to $S_2$ provided every sufficiently small circle centered at $x$ intersects $S$ at exactly two points. Such points are called *ordinary points*. Any curve drawn with a pen or pencil consists primarily of ordinary points. In fact, our everyday experience tells us that almost every point on a curve is an ordinary point, but as with so much else in topology, our everyday experiences are poor guides to mathematical truths.

Points that do not belong to $S_1$ or $S_2$ belong to $S_3$. Points in $S_3$ are called "branch points." A fork in an idealized road is an example of a branch point. One path leads in and two paths lead out, but our curve, $S$, has no preferred direction. Consequently, terms such as "in" and "out" are meaningless, which is why we use the following definition: A point $x$ in $S$ is a branch point of $S$ if all sufficiently small circles centered at $x$ intersect $S$ in at least three points. For some $x$, it might also be true that all sufficiently small circles at $x$ intersect $S$ at more than three points. (Think of an intersection of an idealized road. A circle centered at the intersection would share four points with all sufficiently small circles. Or think of $x$ as the center point of an asterisklike curve. Circles centered at *x* would intersect *S* at many points.) Sierpiński's gasket is a counterexample to the claim that curves must consist primarily of ordinary points. On Sierpiński's gasket, every point is a branch point, or to put it another way, the sets $S_1$ and $S_2$ are empty.

To be clear, the gasket is not a surface. It is more like an idealized net or screen. Place a disc, no matter how small, beneath a Sierpiński gasket, and it will be visible through the gasket. The following is a procedure for creating Sierpiński's gasket:

STEP 1: Imagine an equilateral triangle. Connect the midpoints of each side to form a new equilateral triangle. (It is upside down relative to the first.) Remove the *interior* of the smaller triangle. In other words, the boundary of the triangle remains in place, but all the interior points of the triangle are gone. Now all that remains of the original triangle is three smaller triangles, each with the same orientation as the original large triangle.

STEP 2: Take each of the remaining triangles one at a time. For each such triangle, repeat step 1. (Connect the midpoints of the sides to form another equilateral triangle inside the first and oriented in the opposite direction. Remove the interior points of the inside triangle, but leave the points along the boundary in place.)

STEP 3: At the completion of step 2, there are nine equilateral triangles, each of which is oriented in the same direction as the original. For each of these triangles, repeat the procedure described in step 1.

STEP 4: At the completion of step 3, there are 27 equilateral triangles. Repeat the procedure of step 1 for each of these triangles to produce 81 equilateral triangles. (See the accompanying diagram.) Continue to apply the procedure of step 1 to each triangle.

In the limit, this algorithm removes all of the interior points of all of the triangles, and what remains is a curve. It can be proved that every point of this curve is a branch point except for the points at the vertices of the original (large) triangle. They are regular points. To complete the procedure, perform the entire construction five more times, and join the

*(continues)*

## COUNTEREXAMPLE 4: SIERPIŃSKI'S GASKET
### (continued)

© Infobase Learning

*In this illustration, the white areas are the interiors of the triangles that have been removed. The yellow areas are the points that remain of the original triangle. There are 81 identical triangles remaining. The limit of this process is a curve every point of which is a branch point.*

resulting triangular-shaped curves along their outside edges to form a regular hexagon. This hexagonal latticelike curve has no regular points and no endpoints. It consists entirely of branch points.

A curve consisting entirely of branch points strikes many people as strange, and one might conclude, therefore, that a different definition of the word *curve* is needed to eliminate this "pathology." Subsequent research has shown, however, that no matter how one defines a curve, one can always find a "pathological" example of a curve that satisfies the proposed definition—at least for any definition that has been proposed so far.

*(text continued from page 75)*

Sierpiński constructed a curve with the property that every point on the curve is a branch point. (The curve is described in the sidebar "Sierpiński's Gasket.") How can such a thing exist, or perhaps more to the point, how can such a thing be imagined? It cannot be drawn. No simple formula exists for such a curve. The careful study of set-theoretic topology uncovered a host of such objects.

Early in the history of topology, at a time when it was still called *analysis situs*, mathematicians were motivated primarily by the study of geometric points embedded in spaces—points on the line, points on the plane, and so forth, but recall that even in the 19th century, Ascoli demonstrated that certain sets of functions exhibited some of the same properties as sets of geometric points. Functions could, therefore, also be modeled as sets of points embedded within various "function spaces." The concept of what qualified as a point continued to broaden. As these early topologists gained more experience with examples of spaces, they attempted to identify what they perceived as the fundamental properties of point sets, and they modeled these properties by creating abstract topological spaces with these properties. This is one way to understand what abstract topological spaces are: They are models for classes of more concrete spaces. And because these topologists sought to make their models as general as possible, they discovered many phenomena that were shared by many different-looking spaces.

Topologists have created an entirely new set of ideas, a new mathematical universe, and this new universe is filled with unexpected phenomena. This explains why so many early topologists spent their time searching for examples and counterexamples. It explains why so many of the early articles in topology contained no proofs. Instead, they described unexpected mathematical phenomena. They were of the "Look what I found!" variety of mathematical research. By way of contrast, examples and counterexamples play a very small role in number theory. Number theorists know what it is they study. It took topologists several decades to find out what it was they had created. In some ways their search continues to this day.

# 5

# THE STANDARD AXIOMS
# AND THREE TOPOLOGICAL
# PROPERTIES

What are some examples of topological properties? Why might they be important? As described in the preceding chapter, topology consists of the study of exactly those properties of a topological space that are preserved by the set of all homeomorphisms on that space. From an algebraic point of view, homeomorphisms are very special functions: A homeomorphism must be continuous, it must have an inverse, and its inverse must also be continuous. These are very strong requirements. Most functions are not continuous, and most continuous functions do not have inverses. From a geometric point of view, however, these requirements are very weak. Homeomorphisms generally fail to preserve distances between points, and they may even fail to preserve shapes. Circles, for example, are topologically equivalent to triangles, and triangles are topologically equivalent to hexagons. The set of all points in the plane that are less than one unit from the origin is topologically equivalent to the entire plane, and the entire plane is topologically equivalent to the surface of a sphere with one point removed. When attempting to identify properties that are preserved under homeomorphisms, therefore, our geometric intuition is of limited value. To appreciate topological properties, we need to think about sets of points in new ways.

In this chapter, some fundamental topological properties are described, and we indicate why these properties are important in

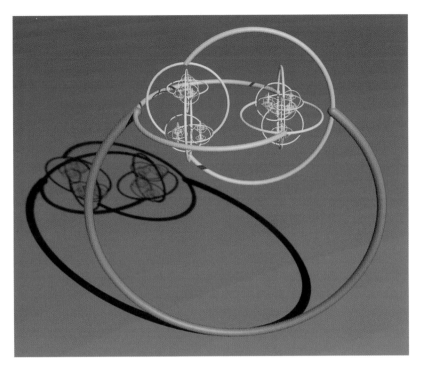

*Alexander's horned sphere. The mathematical set that this sculpture seeks to model is homeomorphic to a ball.*

and out of the field of topology. First, however, we describe what has become a more or less standard definition for a topological space.

## The Standard Axioms

As examples and counterexamples accumulated, topologists reached a consensus about the most efficient and productive way to define a topological space. Despite its simplicity, today's axiomatic definition of a topological space represents the culmination of a great deal of work:

A topological space is a set $X$ together with a collection of subsets of $X$ called a topology. We will call the elements of the topology neighborhoods. The neighborhoods satisfy the following three axioms:

1. The set $X$ and the empty set $\varnothing$ are neighborhoods—that is, they belong to the topology.

2. The union of any (possibly infinite) set of neighborhoods is also a neighborhood.

3. The intersection of any finite set of neighborhoods is also a neighborhood.

These three axioms are often supplemented by a fourth:

4. Given $x_1$ and $x_2$, there exist disjoint neighborhoods $U_1$ and $U_2$ such that $x_1$ belongs to $U_1$ and $x_2$ belongs to $U_2$.

When a space satisfies all four axioms, it is called a *Hausdorff space*. In this volume, all topological spaces are also Hausdorff spaces, but in this section we will concern ourselves only with the three basic axioms that define a topological space. The following are examples of topological spaces:

Example 5.1. Let $X$ represent the real line. We will call a subset $U$ of $X$ a neighborhood if and only if $U$ is the empty set or $U$ is the set $X$ or $U$ is an open subset of $X$, where we use the definition of open subset of the real line that was first described on page 46, chapter 3. (The set of all real numbers and the empty set are also open sets, a fact that we will not show here.) In order to show that the collection of all open subsets of the real line constitutes a topology on the real line, we must show that axioms 2 and 3 are satisfied. Here is one way to do it:

First, we want to show axiom 2 is satisfied—that is, we want to show that the union of any collection of open sets is also an open set. Let $S$ be some collection of open subsets of $X$. The collection $S$ might contain finitely many or infinitely many open sets. We need to show that the union of all the sets in $S$, which we will call $A$, is also an open set.

Let $x$ be a point in $A$, then $x$ must belong to some set $U$ in $S$. The set $U$ is open because every set that belongs to $S$ is open. Because $A$ contains $U$, we conclude that $x$ is an interior point of $A$. Why?

$U$ is an open set containing $x$, and $U$ is also a subset of $A$. This is the definition of an interior point. Since $x$ was an arbitrary point of $A$, we have proved that every point of $A$ is an interior point of $A$. This proves that $A$ is open. Axiom 2 is satisfied.

Second, we must show that the intersection of any finite collection of open sets is open. (This is axiom 3.) Let $S = \{U_1, U_2, U_3, \ldots, U_n\}$ where $n$ represents some natural number and each $U_j$ is open. Let $A$ represent the intersection of all the elements of $S$. We now show that $A$ is open. (We assume that $A$ is not empty.)

Let $x$ be a point of $A$, then $x$ belongs to every set in $S$. Why? Because $A$ consists of exactly those points common to all the sets in $S$. Each $U_j$ is an open set, so $x$ is an interior point of each $U_j$. This means that for every $j$, there is a small interval centered at $x$ that belongs to $U_j$. For each value of $j$, let $I_j$ be an interval centered at $x$ and belonging to $U_j$. Now let $I$ be the shortest of all of these $n$ intervals, then $I$ belongs to every $U_j$, and consequently $I$ belongs to $A$, which is the intersection of all the $U_j$. This shows that $x$ is an interior point of $A$. Since $x$ can represent any point in $A$, we have proved that every point in $A$ is an interior point, and so $A$ is open. Axiom 3 is satisfied. This proves that with the neighborhoods defined as open sets, the real line is a topological space.

Example 5.2. Let $X$ be the interval $\{x: 0 < x < 1\}$. To prove that this is a topological space, repeat the proof in example 5.1 word-for-word.

Example 5.3. Let $X = \{x: 0 < x < 1\} \cup \{x: 2 < x < 3\}$. To prove that this is a topological space, repeat the proof in example 5.1 word-for-word.

Example 5.4. Let $X$ be the points in the plane, and let the set of neighborhoods of $X$ be the set of all open subsets of $X$. (This includes $X$ and $\varnothing$.) Recall that by "open set" we mean that if $U$ is an open set and $x$ is a point that belongs to $U$, then we can draw a small circle centered at $x$ such that all points within the circle belong to $U$. With this definition of neighborhood, $X$ is a topological space. To see that this is true, repeat the proof of Example 5.1 word-for-word except for one small change. Instead of using the word *intervals*, use the word *discs*.

Example 5.5. Let $X$ be the set of all points in the plane that are less than one unit from the origin, and let the neighborhoods of $X$ be the set of all open subsets of $X$. (This includes $X$ and $\varnothing$.) With this definition of neighborhood, $X$ is a topological space. To see that this is true, repeat the proof of Example 5.1 word-for-word except for one small change. Substitute the word *disc* for *interval*.

Example 5.6. The three-point set of Example 4.8 is a topological space when the neighborhoods are defined by the curves.

The modern axioms for a topological space are different from the axioms that Hausdorff used to define neighborhoods. All of the neighborhoods of Examples 5.1 through 5.6, for example, satisfy Hausdorff's axioms as well as the modern axioms for a topological space, but neighborhoods as they are defined in Examples 4.1 through 4.7 fail to satisfy the axioms for a modern topological space. The union of two discs, for example, generally fails to be a disc, and in Example 4.4, neighborhoods were defined as discs. Equipped with the modern definition of a topological space, we now turn our attention to the description of some fundamental topological properties and why they are important.

## Topological Property 1: Compactness

Much of the value of mathematics is that it can be used to describe the world around us. A great deal of effort in engineering and the sciences involves discovering functions that describe physical phenomena. The velocity of air as it flows over a wing, the velocity of water as it flows through a pipe, the employment rate as a function of the price of oil, and the speed and shape of a flame front are all examples of phenomena that are described in terms of mathematical functions. It becomes important in many of these cases to determine the maximum and minimum values of the functions that represent the phenomena of interest. Motivations for finding maximum and minimum values might be to increase efficiency or safety or both. Researchers have developed numerous techniques for finding maximum and minimum values of functions.

It is, however, easy to imagine situations that defeat all of the techniques used to find maximum and minimum values. One need only create functions that have no maximum or minimum values. Consider the function $f(x) = x$ on the interval $\{x: 0 < x < 1\}$. The graph of this function is a line with slope 1. It has neither a maximum nor a minimum value. To see why this is true, consider what happens near zero. The number 0 does not belong to the domain of the function. If we choose any number to the right of zero—we will call the number $x$—then the value of the function, which is also $x$, will be larger than zero. But the number $\frac{x}{2}$ also belongs to the domain of $f$ and $0 < f(\frac{x}{2}) < f(x)$. This demonstrates that the function $f(x) = x$ on the domain $\{x: 0 < x < 1\}$ has no minimum. A similar sort of argument shows that no maximum value exists for this function.

Next, instead of considering the function $f(x) = x$ over the domain $\{x: 0 < x < 1\}$, suppose that we add two points and consider $f(x) = x$ over the new domain $\{x: 0 \leq x \leq 1\}$. The new function has both a maximum and a minimum value. The minimum occurs at $x = 0$, and the maximum occurs at $x = 1$. This is evident, but it is only "evident" because the function is easy to visualize. Often researchers work with functions that are difficult or impossible to visualize, and the search for maximum and minimum values becomes a sort of mathematical exploration. Under these circumstances, it helps to know at the outset whether maximum and minimum values actually exist. In other words, what properties must a continuous function possess in order to ensure the existence of maximum and minimum values? The answer is found in topology.

*Compact sets* constitute one of the most useful classes of sets in set-theoretic topology. Real-valued continuous functions always have maximum and minimum values when their domains are compact. Always. Compact sets are defined in terms of *open covers*, so before we can say what a compact set is, we must explain what an open cover is: Let $X$ be a topological space, and let $C$ be a nonempty subset of $X$. An open cover of $C$, which we will call $S$, is a collection of open sets with the property that every point in $C$ is contained in some (open) element of $S$. Suppose that we write $S = \{U_1, U_2, U_3, \ldots\}$. The set $S$ may be finite or infinite. If it is infinite, it may have uncountably many elements (in which case we would have to use some other set besides the set of natural numbers in the subscript in order to identify the sets). What is important is that each $U_i$ is open. Here are two things to keep in mind about the definition of $S$:

1. It may happen that every point in $C$ belongs to many elements of $S$. The definition of an open cover requires only that every $x$ in $C$ belongs to at least one $U_i$ in $S$.

2. Each set $C$ will have many different open covers. Some open covers may use infinitely many open sets to cover $C$, and some open covers may use only finitely many open sets to cover $C$.

Now suppose that no matter which open cover of $C$ we consider, we can always find a finite subcollection of $S$, which we will call $S'$, such that $S'$ is also an open cover of $C$; then $C$ is called a compact set. To be clear: If for any $S$, $S'$ always exists, then $C$ is compact. If $S$ is already a finite set, just let $S'$ equal $S$. If $S$ is an infinite set, we may have to do some work before we can determine whether $S'$ exists, but as long as every open cover of $C$ contains a finite subcollection that also covers $C$, then $C$ is (as a matter of definition) compact.

Compactness is a topological property. If $A$ and $B$ are topologically equivalent and $f$ is a homeomorphism that transforms $A$ onto $B$, then it transforms compact subsets of $A$ onto compact subsets of $B$, and $f^{-1}$, the inverse of $f$, transforms every compact subset of $B$ onto a compact subset of $A$.

Continuous functions, whether or not they are homeomorphisms, preserve compactness. In particular, if $f$ is a continuous function with a compact domain, then the range of $f$ is compact as well. This is important, and it is a fact that is frequently used in first semester calculus classes. If $f$ is a real-valued continuous function defined on a compact set, then its range is a compact subset of the real numbers, and *every* compact subset of the real numbers contains a largest element and a smallest element. In other words, if $f$ is (1) continuous, (2) real-valued, and (3) has a compact domain, then $f$ attains a maximum and a minimum value. Geometric details about the size, the shape, or even the dimension of the domain are unimportant. All that matters is the topological "structure" of the domain—that is, whether or not it is compact. The following are examples of functions defined on compact domains:

Example 5.7. Let the domain be the interval $\{x: 0 \le x \le 1\}$, and let $f(x) = x^2$. (The graph of $f$ is part of a parabola.) Because the domain is compact and $f$ is continuous, it is guaranteed to have a maximum and a minimum.

Example 5.8. Let the domain be the square $\{(x, y): 0 \le x \le 1, 0 \le y \le 1\}$, and $f(x, y) = x^2 + 2y^2$. This equation can be interpreted as a surface over the square. The function $f$ gives the height of the surface over the square domain at each point of the domain. Because the domain is compact, $f$ is guaranteed to have a maximum and minimum height.

Example 5.9. Let the domain be the cube with edges of length one unit, with sides parallel to the coordinate planes, with one corner at the origin, and lying in the first octant. In symbols, the domain is $\{(x, y, z): 0 \le x \le 1, 0 \le y \le 1, and\ 0 \le z \le 1\}$, and let $f(x, y, z) = x^2 + 2y^2 + 3z^2$. In this case, the coordinates $(x, y, z\ f(x, y, z))$ can be interpreted as the coordinates of a "hypersurface," or four-dimensional surface, over the unit cube. (Admittedly, this is hard to visualize.) Alternatively, $f$ could be interpreted as a function that represents the temperature at each point of the cube, or it could be interpreted as the density of the cube. However we interpret $f$, we can be sure that it has a maximum and a minimum value because its domain is compact.

But how do we know that a domain is compact? The definition of compactness suggests only that we check all open covers of the domain to see if each cover contains a finite subcollection that also covers the domain. This can be a difficult criterion to check. If, however, the domain belongs to the real line, or the plane, or some other finite-dimensional Euclidean space, there is an equivalent criterion that was discovered early in the history of set-theoretic topology. The theorem is named after the French mathematician Emil Borel, who was mentioned in chapter 3, and the German mathematician Heinrich Heine (1821–81). (The theorem was discovered and rediscovered in various forms by several 19th-century mathematicians.) It is one of the most used and useful of all topological theorems. In its original form, the Heine-Borel theorem states that a subset of the real numbers is compact if and only if it is bounded and closed.

To appreciate what the Heine-Borel theorem means, first recall that a set is closed if it contains its limit points, or another way of saying the same thing: A set is closed if its complement is open. If the set is an interval of the real number line, then the set is closed provided it contains its endpoints. In Example 5.7, for example, the domain $\{x: 0 \leq x \leq 1\}$ is closed because the endpoints are included in the definition of the set. We can also show that the domain is closed by verifying that the complement of $\{x: 0 \leq x \leq 1\}$ is open: Choose any number in the complement of $\{x: 0 \leq x \leq 1\}$, and call that number $x_1$. Suppose $x_1$ is greater than 1. A small circle can be drawn about $x_1$ that does not contain 1. The interior of that circle is an open set. Therefore, $x_1$ is an interior point of the complement of $\{x: 0 \leq x \leq 1\}$. A similar argument shows that if $x_1$ lies to the left of 0, then $x_1$ is an interior point of the complement of $\{x: 0 \leq x \leq 1\}$. Because $x_1$ was chosen arbitrarily, this shows that every point in the complement of $\{x: 0 \leq x \leq 1\}$ is an interior point. In other words, the complement of $\{x: 0 \leq x \leq 1\}$ is open. This means that $\{x: 0 \leq x \leq 1\}$ is closed. (This type of indirect argument is common in set-theoretic topology.)

Second, a set of real numbers is bounded if it is contained within an interval of finite length. As already mentioned in chapter 3, another way of thinking about this criterion is to imagine drawing a circle centered at the origin. If a circle can be drawn so that it contains the set of interest, then the set is bounded. It might be

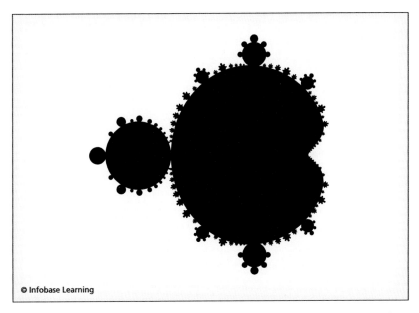

© Infobase Learning

*If the white part of the diagram is an open set, then by the Heine-Borel theorem the black part of the diagram is compact.*

necessary to draw a very large circle. The size of the circle does not matter. It is important only that some such circle exists.

The Heine-Borel theorem applied to the real line provides a simpler set of criteria to test whether a subset of the real numbers is compact. If a subset of the real numbers is closed and bounded, then it is compact, and if it is compact, it is closed and bounded. When working on the real line, we can avoid considering open covers and concentrate on whether the set is closed and bounded, and this is usually much easier to verify than the open cover criterion.

Today the Heine-Borel theorem is usually stated in terms of finite-dimensional Euclidean spaces, one example of which is the real line. A more modern definition of the Heine-Borel theorem states that a subset of $n$-dimensional Euclidean space is compact if and only if it is closed and bounded.

Example 5.10. The set $\{x: 0 \leq x \leq 1\}$, which was considered in Example 5.7, is closed and bounded and hence compact. By way of contrast, consider the set $\{x: 0 < x < 1\}$. This set cannot be compact because it is not closed. (To see that it is not closed, notice that it does not contain the numbers 0 and 1, which are limit points of the set. A closed set always contains all its limit points.)

Example 5.11. The unit square, which is the domain described in Example 5.8, is compact. Here is why: First, the set is bounded. The point on the square that is farthest from the origin is the corner point (1, 1). It is $\sqrt{2}$ units away from the origin. Therefore, any circle centered at the origin and with a radius greater than $\sqrt{2}$ units will contain the square. Second, the unit square is closed because, as a matter of definition, it contains its boundary. Having demonstrated that the square is closed and bounded, we conclude, with the help of the Heine-Borel theorem, that the square is compact.

(Here is a very brief alternative proof of the compactness of the square: As pointed out earlier, a continuous function transforms a compact set onto a compact set. The interval $\{x: 0 \leq x \leq 1\}$ is compact. Peano devised a continuous function with domain equal to $\{x: 0 \leq x \leq 1\}$ and range equal to the unit square. Therefore, the unit square is compact.)

Example 5.12. To see that the unit cube, which is the domain described in Example 5.9, is compact, repeat what is said in Example 5.11, with the following changes: The point farthest from the origin is (1, 1, 1), not (1, 1), and its distance from the origin is $\sqrt{3}$, not $\sqrt{2}$. Substitute the words *sphere* for *circle* and *cube* for *square*. It is that easy. With the help of the Heine-Borel theorem, we conclude that the unit cube is compact.

Compactness is a very convenient property for a set to have. Mathematicians have often made their work easier by requiring that the sets with which they work be compact because compactness often leads to great simplifications. This was recognized as early as the 19th century, well before the word *compact* was coined.

# Topological Property 2: Regularity

Some of the spaces that topologists have studied were created by topologists to illustrate a particular property or to solve a particular problem. Other times, new-looking spaces have been created by researchers from outside the field of set-theoretic topology. The creation of such spaces is often motivated by practical concerns. Engineers, scientists, and mathematicians have all created spaces for their own uses, but it has not always been apparent whether these spaces were really new or just new looking. In other words, from a mathematical point of view, what did these researchers do? To a topologist, the question is important because part of the job of a topologist is to attempt to understand the logical relationships that exist among topological spaces. They seek to establish a sort of taxonomy of topological spaces; they want to know the many ways that different-looking spaces are related to each other. Topologists engaged in this effort are attempting to bring order to chaos.

One early illustration of the advantage of paying careful attention to foundational questions occurred early in the 20th century. In 1905, the American mathematician Oswald Veblen (1880–1960) produced a proof of the Jordan curve theorem, a famous result that states that a simple closed curve in the plane divides the plane into exactly two disjoint regions, an inside and an outside. (A curve is called *closed* if it has no endpoints and is called *simple* if it does not cross itself. A circle is an example of a simple closed curve.) The Jordan curve theorem is famous in part because it seems so obvious and yet turned out to be so difficult to prove. The theorem had been proved before Veblen wrote his paper, but the original proof depended on the existence of a metric, a function that enables one to compute the distance between two points. Veblen's proof was based on a set of axioms that made no mention of a metric. He concluded that his proof was "purely topological," by which he meant that it did not depend on the concept of distance. Veblen claimed that he had done something new, but had he done what he claimed?

The American topologist Robert L. Moore showed that although Veblen did not make explicit use of the concept of distance, the existence of a metric was a logical consequence of the axioms on which Veblen's proof was based. In other words, Veblen's work looked new—it seemed as if he had demonstrated that the Jordan curve theorem could be proved in a space without a metric—but

## THE ROLE OF RIGOR IN MATHEMATICS

Perhaps more than most mathematicians, topologists have been interested in sets of axioms and in the rigorous application of the axiomatic method. They pare their axioms down to a minimum in order to develop as broad a theory as possible, and then they add additional axioms, often one at a time, to determine which properties of a space are dependent on which axioms. This is mathematics as a branch of logic.

Not everyone is satisfied with this approach to mathematics. Many mathematicians were, after all, quite content to develop calculus as a series of algorithms without worrying overly much about the logical foundations of the subject. The efforts of Bolzano, Dedekind, Cantor, and Weierstrass to place mathematics on a firm logical foundation were also criticized by many of their contemporaries. Not everyone agreed on the necessity of placing "pathological functions" and logical paradoxes at the center of mathematical thought. For many of the critics, mathematics is a conceptual tool that is of value only to the extent that it facilitates progress in engineering and science.

The British scientist, engineer, and mathematician Oliver Heaviside (1850–1925) is an excellent example of this attitude. Heaviside was one of the most successful scientists and engineers of the 19th century. He made important contributions to the theory of electromagnetic phenomena, including the propagation of radio waves and the design of

*Oliver Heaviside believed that mathematics should be grounded in experiment, just like science and engineering.* (Dibner Library of the History of Science and Technology)

Moore proved that Veblen had created a *metric space;* he just did not realize it. Veblen had done nothing new. His different-looking proof was logically equivalent to earlier established proofs.

electrical circuits. He was also an extraordinarily creative mathematician. In order to complete some of his engineering and scientific work, he created an "operational calculus," a new branch of mathematics. He used the operational calculus to solve certain equations that had arisen in his research. Although he undoubtedly considered himself a scientist and engineer, the operational calculus is sometimes described as Heaviside's greatest accomplishment. Even so, he rarely missed an opportunity to express his disdain for classical proof-oriented mathematics and the mathematicians who study it. Here is a quote from his book *Electromagnetic Theory,* volume 3, in which he offers a few opinions on the teaching of geometry and the role of proof:

> Euclid is the worst. It is shocking that young people should be addling their brains over mere logical subtleties, trying to understand the proof of one obvious fact in terms of something equally, or, it may be, not quite so obvious, and conceiving a profound dislike for mathematics, when they might be learning geometry, a most important fundamental subject, which can be made very interesting and instructive. I hold the view that it is essentially an experimental science, like any other, and should be taught observationally, descriptively, and experimentally in the first place.

The debate did not end with Heaviside. Today some mathematicians emphasize that the axiomatic method has its own shortcomings, and ever more powerful computers enable mathematicians to investigate mathematical phenomena in entirely new ways. To these researchers, more emphasis should be placed on mathematics as an experimental discipline. They are, although they may not know it, disciples of Oliver Heaviside. Not only should questions be investigated computationally, they argue, but numerical experiments should be given the same degree of respect in mathematics as experiments are given in the physical sciences, in which everyone acknowledges that experimental results are the bedrock upon which all scientific knowledge rests. While classical mathematical arguments remain extremely important to mathematical progress, experimental mathematics is growing in importance, calling into question what it means to do mathematics.

One problem to which topologists have devoted a great deal of attention is the identification of conditions that are necessary and sufficient to ensure the existence of a metric, which is also called a distance function. A metric enables one to talk about the distance between points. Keep in mind that the spaces are abstract. The points might be functions, or functions of functions, or geometric points, or something else entirely. Topologists want to know what properties a topological space must have in order to ensure that a metric exists. They seldom are concerned with finding a formula for the metric. They only want to know that a formula could, in theory, be found. Experience has shown that even this question is not an easy one to answer, but before we look at one very important answer to this question, we need to specify exactly what is meant by a metric.

Let $X$ be a topological space. A metric on $X$ is a function of two variables. It is usually written $d(x, y)$, where $x$ and $y$ belong to $X$. The value of $d(x, y)$ is interpreted as the distance between the points $x$ and $y$. In order that the interpretation of "distance between points" make sense, $d(x, y)$ must satisfy three criteria:

1. $d(x, y) \geq 0$ for all $x$ and $y$ in $X$

2. $d(x, y) = 0$ if and only if $x = y$

3. For any three points $x$, $y$, and $z$, it is always true that
   $d(x, y) \leq d(x, z) + d(z, y)$.

Property 3 is called the *triangle inequality*. It can be interpreted as a statement that the length of one side of a triangle is never longer than the sum of the lengths of the two remaining sides. (The points $x$, $y$, and $z$ can be interpreted as the locations of the vertices of the triangle.)

Example 5.13. Let $(x_1, y_1)$ and $(x_2, y_2)$ be points in the plane. Define $d((x_1, y_1), (x_2, y_2)) = \sqrt{(x_1 - x_2)^2 + (y_1 - y_2)^2}$. Usually called "the" distance formula, this is a metric on the plane. Mathematicians sometimes use metrics different from this one when studying the plane.

Metric spaces—those topological spaces on which a metric can be defined—have many properties in common. Many of these properties have been identified and exhaustively studied. Knowing that a topological space is a metric space reveals a lot about the structure of the space, but not every topological space can accommodate a metric. Or, to put it another way, sometimes it is not possible to define a metric on a topological space. Whether a metric can be defined on a particular topological space is solely a function of the topological properties of the space. The Russian topologist Pavel Samuilovich Urysohn (1898–1924) discovered conditions sufficient to ensure that an abstract topological space can support a metric.

Urysohn entered the University of Moscow in 1915 to study physics, but his interest shifted to mathematics after he came into contact with the mathematics faculty. Russia has long produced some of the world's most successful mathematicians, and many of them have attended or taught at the University of Moscow. Urysohn did both. He graduated in 1919 and remained at the university to obtain a Ph.D., which he received in 1921. While at the university, he became friends with Pavel Sergeevich Aleksandrov (1896–1982), who would also become a successful topologist.

Soon after graduation, Urysohn made several important contributions to dimension theory and also discovered what is now known as the Urysohn metrization theorem. Within a few years, he and Aleksandrov went on a tour of Europe, meeting some of the most influential mathematicians of the day, including David Hilbert, who did research in almost every type of mathematics, and the Dutch philosopher and topologist Luitzen E. J. Brouwer. Unknown to Urysohn, Brouwer had also published a theory of dimension. It was different from that of Urysohn's, and upon learning of Brouwer's work, Urysohn quickly identified a flaw in Brouwer's logic, a feat that favorably impressed Brouwer. Urysohn and Aleksandrov soon left Amsterdam and continued their journey westward across the continent, seeing the sights and meeting mathematicians. In France, while swimming in the Atlantic, Urysohn drowned.

To appreciate Urysohn's metrization theorem, one needs to know three definitions. The first is the definition of a *basis* for a topological space $X$. A basis is a collection of subsets of $X$. The concept of a basis was invented because often it is too difficult to explicitly specify a topology. It is often far easier to specify a basis. The basis is then used to describe the topology. A basis is, therefore, "almost" a topology. A basis for a topology on $X$ is a collection of subsets of $X$ that satisfies two properties:

1. Every point in $X$ belongs to at least one basis element.

2. If a point $x$ belongs to the intersection of two basis elements, which we will call $B_1$ and $B_2$, then there exists a third basis element, which we will call $B_3$, that contains $x$ and belongs to the intersection of $B_1$ and $B_2$. (This is also Hausdorff's second axiom. See chapter 4.)

A subset $U$ of $X$ is open relative to a particular basis if for every $x$ in $U$ there exists a basis element $B$ such that $x$ belongs to $B$ and $B$ belongs to $U$. Specifying a basis is, therefore, just a simpler method of specifying the open sets in $X$.

Example 5.14. The set of all open intervals on the real line forms a basis for the standard topology of the real line.

The second thing one must know before describing Urysohn's metrization theorem is what it means to speak of the boundary of a set. Let $X$ be any topological space, and let $U$ be a subset of $X$. A point $x$ is a boundary point of $U$ if every open set that contains $x$ contains elements of $U$ and elements of the complement of $U$. It does not matter whether $x$ belongs to $U$. By way of example, the boundary of the set $\{x: a < x < b\}$ is the set $\{a, b\}$, and $\{a, b\}$ is also the boundary of the set $\{x: a \leq x \leq b\}$.

The third, and last, concept is that of a *regular space*. Let $X$ be a Hausdorff space. (The Hausdorff condition ensures that sets consisting of one point are closed.) Let $x$ be any point in $X$, and let $V$ be an open set containing $x$. The space $X$ is regular provided there always exists an open set $U$ containing $x$ such that $U$ and its boundary belong to $V$. (See, for example, the diagram on page 59, example 4.1, which illustrates that the real line is a regular space.) The property of being a regular space is a topological property—in other words, it is preserved under homeomorphisms.

It is easy to imagine sets of geometric points that satisfy the definition of regularity. In fact, it can be difficult for someone who has not studied topology to imagine a topological space that fails to satisfy the definition of regularity. Essentially, regularity is a requirement that a closed set and a point that does not belong to it can be separated. In the definition of the preceding paragraph, $x$ is the point, and the closed set is the complement of the open set $V$. The set $U$ contains $x$ without impinging on the complement of $V$, and it is in this sense that $x$ and the complement of $V$ are separated. When a topologist creates a topological space, he or she may supplement the three basic axioms that define a modern topological space with the regularity axiom. Other times, topologists may be given a topological space and then attempt to determine whether the space is also regular.

Urysohn proved that if a topological space has a countable basis—that is, if the sets that constitute the basis can be put into one-to-one correspondence with the natural numbers—and if the topological space is regular, then a metric can be defined on the space.

Urysohn's metrization theorem is a classical result in set-theoretic topology. It is important because a metric is a very convenient property for a topological space to have, and Urysohn's theorem often enables mathematicians to determine whether a topological space has a metric. In that sense, it is a powerful theorem, but the properties with which the theorem is concerned—countable bases, regularity, and the existence of a metric—are outside the everyday experience of almost everyone. Say what one will about Euclid, almost everyone can draw a triangle. Say what one will about Urysohn, almost no one can draw an existential metric. Topology, because the concepts are so primitive, is concerned with ideas that require substantial preparation just to state. This is one reason why, despite its great value within mathematics, topology has so far failed to attract much attention outside mathematics.

# Topological Property 3: Connectedness

One of the best-known and most fundamental theorems of calculus is called the intermediate value theorem. It is usually encountered in a first semester calculus course, and most students assume its truth long before they take a course in calculus. It concerns continuous real-valued functions defined over compact subsets of the real line. It is described as follows:

> The intermediate value theorem: Let $f$ be a continuous real-valued function with domain $\{x: a \leq x \leq b\}$. If $r$ is any real number between $f(a)$ and $f(b)$, there is a real number $c$ between $a$ and $b$ such that $f(c) = r$.

In other words, as the independent variable $x$ assumes every value from $a$ to $b$, the dependent variable $f(x)$ must assume every value from $f(a)$ to $f(b)$. Another rough way to summarize the intermediate value theorem is to say that if the domain of the function has no gaps in it, neither does its range. A great many results in analytic geometry and calculus depend on the truth of the inter-

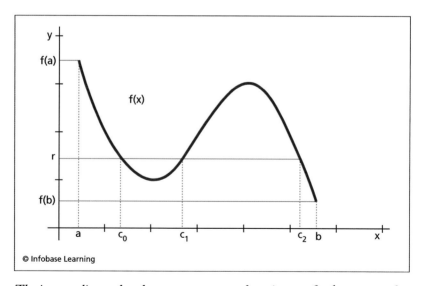

© Infobase Learning

*The intermediate value theorem guarantees the existence of at least one value of c such that (when the hypotheses of the theorem are satisfied) f(c) = r.*

mediate value theorem, and the truth of the theorem depends on the topological notion of *connectedness*.

As with most topological concepts, some care is required to define connectedness in order to obtain a definition that is logically coherent instead of merely plausible. This requires the twin notions of *separation* and connectedness. Let $X$ be a topological space. A separation of $X$ is accomplished provided one can find two nonempty open disjoint sets, $U$ and $V$, such that $X = U \cup V$. In other words, a separation of $X$ must satisfy the following three conditions:

1. Every point in $X$ is an interior point of $U$ or an interior point of $V$.

2. No point belongs to both $U$ and $V$.

3. There is at least one point of $X$ that belongs to $U$ and one point of $X$ that belongs to $V$.

Sometimes it is easy to effect a separation of a topological space. Consider the topological space $X = \{x: 0 < x < 1\} \cup \{x: 2 < x < 3\}$, which was described in Example 5.3. Let $U = \{x: 0 < x < 1\}$ and $V = \{x: 2 < x < 3\}$. This constitutes a separation of $X$. Other times, a topological space cannot be separated, in which case it is called a connected space. The set $\{x: 2 < x < 3\}$, for example, cannot be separated and is, therefore, connected.

When a space can be separated, one can quickly deduce a result that many people find surprising—at least at first. Recall that closed sets were first defined in terms of open sets: A subset of a topological space is closed if its complement is open. This use of the words *open* and *closed* is peculiar to topology. It has nothing to do with common notions of open and closed. In ordinary speech, *closed* and *open* are antonyms. As with *light* and *heavy* and *near* and *far*, *closed* and *open* are opposing notions in ordinary speech. A door, for example, cannot be simultaneously closed and open. In topology, however, sets can be simultaneously closed and open. This situation occurs when a topological space can be separated.

To see how sets in a topological space can be open and closed, consider, again, the topological space $X = \{x: 0 < x < 1\} \cup \{x: 2 < x < 3\}$ first encountered in Example 5.3. As noted earlier, if we define $U = \{x: 0 < x < 1\}$ and $V = \{x: 2 < x < 3\}$, then we have a separation of $X$. The set $U$ is open because every point in $U$ is an interior point. Therefore, the complement of $U$, which is $V$, must, as a matter of definition, be closed, but it is also true that every point in $V$ is an interior point. Therefore, $V$ is also open. Now we repeat the preceding sentences interchanging the role of the sets $U$ and $V$: The set $V$ is open because every point in $V$ is an interior point. The complement of $V$, which is $U$, must, as a matter of definition, be closed. We can only conclude that $U$ is open and closed and that $V$ is open and closed.

The property of connectedness is a topological invariant. If a topological space is connected, then this property is preserved by all homeomorphisms. It is, in fact, preserved by all continuous functions. The topological space $\{x: a \leq x \leq b\}$, which is the domain of the function in the intermediate value theorem, is a connected topological space. Since the domain of $f$ is connected, the range of $f$ is also connected. Therefore, if $r$ lies between $f(a)$ and $f(b)$, then $r$ lies in the range of $f$. Otherwise, we could separate the range of $f$ into the set of points less than $r$ and the set of points greater than $r$. Because we cannot separate the range of $f$, we can conclude that there is an element in the domain, which we will call $c$, such that $f(c) = r$.

When topology is described in newspapers and magazines as "rubber sheet geometry," part of what the writer is attempting to convey is the concept of connectedness. Connected spaces that are stretched, compressed, or otherwise continuously deformed remain connected. Connectedness, when applied to the real line or the plane, is, at least initially, not too difficult to understand, and mathematicians frequently assumed that sets were connected long before topologists undertook the study of connected spaces. As with most concepts in topology, the idea arose from the study of geometric spaces such as the line and the plane. Over time, the definition was revised until it was general enough to apply to topological spaces that have no easy geometric interpretation, but the revisions were made so as to keep the definition relevant to the simpler (geometric) spaces in which the concept first arose.

The concepts of compactness, regularity, and connectedness are fundamental to the study of topological spaces. With respect to the line or the plane, these concepts are fairly straightforward because they are founded in geometry. Precise definitions are, however, complicated by the desire of topologists to develop a theory that is general enough to account for topological spaces that cannot be interpreted geometrically. It is in this sense that topological concepts go beyond geometry to reveal new and important properties of general topological spaces. In his book *The Foundations of Science*, Henri Poincaré described *analysis situs*, the original term for topology, and why he felt that its increased level of abstraction was so important:

> There is a science called *analysis situs* and which has for its object the study of the positional relations of the different elements of a figure, apart from their sizes. This geometry is purely qualitative; its theorems would remain true if the figures, instead of being exact, were roughly imitated by a child. We may also make an *analysis situs* of more than three dimensions. The importance of *analysis situs* is enormous and cannot be too much emphasized; . . . We must achieve its complete construction in the higher (dimensional) spaces; then we shall have an instrument which will enable us really to see in hyperspace and supplement our senses.

Poincaré wrote these words before topology had established itself as an independent branch of mathematics. Many modern topologists have more inclusive ideas about the nature of topology and its uses, but the characterization of topology as a tool that enables one to see into hyperspace is still very relevant.

# 6

## SCHOOLS OF TOPOLOGY

From the late 19th century through much of the 20th century, set-theoretic topology developed at a furious rate. New ideas were regularly introduced, new examples and counterexamples were discovered, and new theorems were proved. During this time, many mathematicians who did not specialize in set-theoretic topology nevertheless followed the development of the subject because progress in their disciplines depended on progress in set-theoretic topology. Today, set-theoretic topology is a "mature" branch of mathematics. Progress has slowed—at least temporarily. Unresolved questions remain. Mathematicians continue to do research in the field, but more recent discoveries have not had the same impact that the older discoveries had. As with several other very important branches of mathematics, including linear algebra, measure theory, and even Euclidean geometry, the older discoveries remain at the center of mathematical thought, and the newer discoveries are primarily of interest to topologists. (In the 1980s, Paul Erdös, one of the 20th century's most prolific mathematicians, gave a series of lectures that consisted of listing in rapid succession numerous unsolved problems in Euclidean geometry. Even after 2,000 years, he remarked, a number of interesting and unsolved problems remained, and he hoped that he could inspire further research into the geometry of Euclid.) It is a testimony to their success that topologists were able to develop their subject so thoroughly in so short a time.

Although early topological research was vigorously pursued by mathematicians from around the world, those in some regions contributed more to the development of the subject than others.

The former Soviet Union and what is now the Czech Republic, for example, were important centers of research in set-theoretic topology. They still are, but three cities in particular are closely associated with the development of the subject. They are Warsaw, Poland; Austin, Texas; and Tokyo, Japan. At universities in each of these cities, a very distinct "school" of mathematical thought developed. In addition to contributing to progress in topology, the work that was carried on in these institutions influenced the broader development of mathematics.

## The Polish School

To better appreciate the Polish contribution to topology, it helps to know a little about the history of Poland. As an ethnic group, the Polish people have their own distinct language and social customs, and they have long occupied a large area near the center of the European continent. Even so, they have not always lived in the nation of Poland. The political history of Poland is a volatile one, and the borders of the Polish state have expanded and contracted repeatedly over the centuries. At the beginning of the 20th century, Poland as an independent political entity did not exist: Russia, Germany, and the Austro-Hungarian Empire each governed part of the region occupied by the Polish people.

At the end of World War I, Poland was reconstituted as a nation. Feelings of nationalism were running high. One Polish mathematician in particular, Zygmunt Janiszewski (1888–1920), wanted to establish a Polish school of mathematics. He wanted to put the newly reestablished nation on an equal footing with other nations. Mathematics was an excellent place to start. Mathematical research does not require expensive research facilities, and this was an important consideration since these facilities did not exist in Poland at the time. Just as important, there were already a number of first-rate Polish mathematicians, and if they concentrated their efforts on a few rapidly growing branches of mathematics, they could quickly make their presence felt throughout the world. Cooperation was necessary, Janiszewski believed, because the number of available mathematicians was not large enough to sustain serious research programs in all the major branches of mathematics. Set-theoretic topology was one of the branches of mathematics that they chose, and the University of Warsaw became the center for research in set-theoretic topology in Poland. Other Polish universities later contributed to the effort.

Soon, these mathematicians founded a professional society to facilitate the exchange of information. They also established a research journal, *Fundamenta Mathematicae*, through which they could make their discoveries known. (As was mentioned in chapter 4, after he was banned from publishing in Germany, Hausdorff continued to publish in *Fundamenta Mathematicae*.) They began to contribute to the development of topology in important ways. Janiszewski witnessed only the beginning of the research program for which he was responsible. He died in the global influenza pandemic that began near the end of World War I.

Topology flourished in Poland between the two world wars. Some of the more successful members of what came to be known as the "Polish school" of topology are Karol Borsuk (1905–82), Bronisław Knaster (1893–1980), Kazimierz Kuratowski (1896–1980), Stefan Mazurkiewicz (1888–1945), Juliusz Schauder (1899–1943), and Wacław Sierpiński, the creator of the Sierpiński gasket (See the sidebar "The Sierpiński Gasket" in chapter 4 on pages 76–78.)

*Coin commemorating Wacław Sierpiński's time as president of the Warsaw Scientific Society (T.N.W.). Sierpiński published more than 700 articles on mathematics and authored numerous books. (C. Szczebrzeszynski)*

Mathematical research slowed during World War II. During the German occupation of Poland, the universities were closed, and teaching was forbidden. As a matter of policy, academics were sometimes executed. Borsuk and Kuratowski continued to teach in secret. Borsuk was captured and briefly imprisoned for his activities. He escaped and remained in hiding for the remainder of the war, but he found a way to smuggle his research papers out of Poland to be published elsewhere. Kuratowski, who was Jewish, also collaborated with the Polish resistance movement. Mazurkiewicz, weakened by privation and disease, died shortly after the end of the war. Schauder, who was in Lvov when it was captured by the German army, died soon after the onset of the occupation under uncertain circumstances. (At the time, Lvov was part of Poland. It is now part of Ukraine.) At the conclusion of the war, all of the surviving mathematicians resumed their research activities.

There are too many mathematicians in the Polish school of topology to describe all of their activities here, but one of the most interesting of these mathematicians was Kazimierz Kuratowski. We focus on just one aspect of his work that is both accessible and revealing. During the interwar years Kuratowski published a two-volume treatise on topology. This work is notable not only for its content but for the way Kuratowski expressed his ideas. He used algebra. Set-theoretic topology is, even today, often expressed without much in the way of symbolic notation. It is unusual in this regard. In fact, in many branches of mathematics, algebraic notation is so pervasive that it is doubtful that the mathematical ideas characteristic of these subjects could be expressed without the help of algebraic notation. Contrast that with, for example, the standard topological axioms, which are listed at the beginning of chapter 5. They are more representative of how ideas are expressed in topology, and they contain little in the way of specialized symbolic notation. Axiom 2, for example, states, "The union of any (possibly infinite) subcollection of neighborhoods is also a neighborhood." That topology contains so little in the way of specialized notation is surprising, given the fact that algebraic notation has proven so useful elsewhere. Symbolic notation makes it much easier to express complex mathematical ideas clearly and unambiguously. It was, after all, created for just that purpose.

Topology can incorporate algebraic symbolism. In his treatise, Kuratowski illustrated how one can use algebraic notation to express set-theoretic topology. To understand how this works, keep in mind that topologies are determined by the way the neighborhoods are defined. Neighborhoods, not individual points, are what matter. They determine the topological structure of the parent set. In fact, in topology, the word *point* conveys very little information at all. Although the word *point* originally had a specific meaning—it originally meant a "geometric point"—by the early 20th century, the word was simply a placeholder, an undefined concept. A point in topology is just the raw material from which the neighborhoods are constructed. By itself, the term is meaningless.

To create a symbolic language for topology, Kuratowski turned toward the work of the 19th-century British mathematician George Boole (1815–64), who today is best remembered for his pioneering work in creating what is now known as Boolean algebra. Boole sought to express logic as a sort of algebra. (Boolean algebra is now used in the design of logic circuits in computers.) Boole interpreted his algebra in terms of sets and set theoretic operations. In Boole's algebra, the operation of intersection, which is represented by the symbol $\cap$, corresponds to the operation of multiplication. The operation of union, which is represented by the symbol $\cup$, corresponds to addition. With these correspondences, it is easy to show, for example, that intersection distributes over union—in other words, $A \cap (B \cup C) = (A \cap B) \cup (A \cap C)$ for any sets $A$, $B$, and $C$—just as multiplication distributes over addition in the algebra we first learn in junior high and high school. However, in contrast with high school algebra, in Boole's algebra, $x^2 = x$ (that is, for any set, $A \cap A = A$), because the intersection of a set with itself is just the original set.

Kuratowski adopted Boolean algebra to express relationships between neighborhoods. Boole's algebra is not quite adequate for the purpose for which Kuratowski had in mind because Boole had no concept of topology, but Kuratowski discovered that it was almost adequate. He supplemented Boolean algebra with the *closure* operation. The closure of a set $A$, which is written $\overline{A}$, is the set consisting of all the elements of $A$ and all the limit points of $A$. (By way of example, the closure of $\{x: 0 < x < 1\}$ is the set $\{x: 0 \leq x \leq 1\}$, and the closure of the set $\{x: 0 \leq x \leq 1\}$ is just itself.)

With this notation, Kuratowski was able to express topological theorems and their proofs using algebraic notation. This was an important innovation, and it allowed him to gain new insights into set-theoretic topology. The notation is terse, but once a reader becomes accustomed to it, its superiority over conventional topological language is clear. In Kuratowski's notation, as in Boole's notation, the parent set is denoted by the number 1, and the empty set is denoted by the number 0. In other words, whereas we have been writing $X$ to represent the parent set, Kuratowski wrote 1. The symbol $\subset$ means "is a subset of." Here are Kuratowski's axioms for a topological space. (The letters $X$ and $Y$ denote arbitrary subsets of 1.)

Axiom 1: $\overline{X \cup Y} = \overline{X} \cup \overline{Y}$

Axiom 2: $X \subset \overline{X}$

Axiom 3: $\overline{0} = 0$

Axiom 4: $\overline{\overline{X}} = \overline{X}$

Axiom 1 says that the closure of the union of two sets equals the union of the closures, axiom 2 says that every set belongs to its own closure, axiom 3 says that the closure of the empty set is the empty set, and axiom 4 says that the closure of the closure of a set equals the closure of the set. These axioms form the foundation for his exposition of topology. Volume I is more than 500 pages long and full of interesting mathematics. It is a remarkable accomplishment.

Since it was first published in the 1930s, Kuratowski's treatise has been translated, updated, and published multiple times. It is a classic. It is also an early example of set-theoretic topology turning inward in the sense that the results that topologists began to generate—as well as the language in which those results were expressed—became increasingly inaccessible to nontopologists. In fact, Kuratowski's notation has not been widely adopted. Certainly, some topologists have found it convenient to use, but it is hard to find a mathematician who specializes in analysis, for example, who can easily read it, even though analysis, that branch of mathematics that grew out of calculus, is heavily dependent on results from topology. This continual drive toward specialization is common in mathematics. As a mathematical discipline develops, specialists begin to aim their research at an ever narrower audience, an audience that soon consists of a relative handful of other specialists. Anyone who has attended a few mathematics seminars at a college or university has noticed that some in the audience bring papers to correct during the lecture, secure in the knowledge that whatever is happening at the front of the room is irrelevant to their area of research and perhaps incomprehensible as well.

There are some who claim that mathematics is no longer a single discipline—and that it has not been a single branch of knowledge for many years—a result of the trend toward increased specialization. Papers in topology, algebra, and analysis are now written for topologists, algebraists, and analysts, respectively, which is why some say that mathematics has fragmented into a jumble of niche areas, each of which is unintelligible to researchers in other mathematical niches. There is some truth to this assertion, but the situation is more complex. On the one hand, part of what it means to become an expert is that one develops an appreciation for ideas and techniques that are not known (or at least not appreciated) outside of one's area of expertise. On the other hand, no matter how different two mathematical disciplines appear to be, both are examples of the axiomatic method put into practice, and most branches of mathematics are still built on common algebraic and/ or topological concepts. To the extent that mathematics remains unified, those unifying concepts are to be found in topology and algebra, and with effort on the part of both the presenter and the audience, some of what is important about mathematical research is sometimes communicated to people who are not experts in the field. Most would agree that communication of this sort takes real effort on both sides.

## Topology at the University of Texas

Topological research at the University of Texas is firmly connected with the American mathematician Robert Lee Moore (1882–1974).

*Faculty and staff of the department of mathematics at the University of Texas late in Robert Moore's tenure. (Moore is in the fourth row up, second from the left.) During Moore's time at the university, the math department was known as much for Moore's innovative method of teaching as for the advanced research that was (and still is) performed there.* (Robert E. Greenwood Papers, 1881–1999, Archives of American Mathematics, Dolph Briscoe Center for American History, University of Texas at Austin)

Moore was born in Dallas, Texas, and obtained his bachelors degree from the University of Texas at Austin. After receiving his degree, he remained at the university for one year. During this time, he discovered that a collection of axioms devised by the German mathematician David Hilbert, who was one of the most influential mathematicians in the world at this time, were not entirely logically independent. Part of one of the axioms was actually a logical consequence of the others. It was a remarkable discovery for someone with only a modest background in higher mathematics.

Word of Moore's discovery eventually reached Eliakim Moore (no relation), a professor of mathematics at the University of Chicago, who then arranged for Robert Moore to pursue graduate education at the University of Chicago. When the invitation came, Robert Moore was working as a high school mathematics teacher in Texas. Moore eventually obtained a Ph.D. in mathematics from the University of Chicago, and for a number of years thereafter he moved about the country, teaching at different universities, developing his ideas about topology, and honing an approach to teaching that continues to inspire debate about what a mathematics education should emphasize and how best to educate students in mathematics. (See the sidebar "The Moore Method.") To Moore's great delight, he eventually joined the faculty of his alma mater, the University of Texas, and he remained at the university for the remainder of his working life.

For Moore, ". . . the primary question was not 'What do we know?' but 'How do we know it?'" Those words do not, however, refer to Moore. They are taken from the Greek philosopher Aristotle's description of Thales of Miletus, whom the Greeks recognized as the first mathematician in their mathematical tradition. Nevertheless, the remark applies just as well to Moore, who spent his life studying the foundations of set-theoretic topology and set theory in general. The work for which Moore is best known is called *Foundations of Point Set Theory*. The way that he presents the subject demonstrates his interest both in topological research and in the logical structure of topology itself. In particular, he spent a great deal of time making explicit various relationships among specific theorems and specific axioms.

In the first four chapters of the seven-chapter *Foundations of Point Set Theory*, Moore develops a conception of topology that is based on six axioms. Moore numbers them 0, 1, 2, 3, 4, and 5. The first chapter is concerned exclusively with developing the logical consequences of axioms 0 and 1. In the second chapter, he proves theorems based on axioms 0, 1, and 2. In the third chapter, he proves theorems based on axioms 0 through 4, and in the fourth chapter, he considers the logical consequences of the full set of axioms. In this way, he clearly shows how various topological theorems depend on specific axioms. Most mathematicians aspire to develop their mathematics in a rigorous way, but few pursued the goal with as much determination as Moore.

Early in his career, Moore spent a lot of his time identifying relationships between general topological spaces and metric spaces. He also spent a lot of time "fine tuning" various sets of axioms, changing a word here and there and investigating the logical implications of each change. (During the 1920s and 1930s, when Moore did much of his research, there was no generally agreed upon definition of a topological space, and researchers frequently investigated competing sets of axioms in order to determine which set yielded better—"better" in the sense that they were conceptually more appealing—results.) Central to these concerns is, of course, the concept of neighborhood.

From a narrow point of view, the precise way in which neighborhoods are specified is unimportant. As long as the resulting collection of neighborhoods satisfies the axioms that define what a topological space is, the parent set together with its neighborhoods constitute a topological space, and any such space can be studied on its own terms. From a broader point of view, given a set $X$, two different-looking definitions for the neighborhoods of $X$ can sometimes result in the same topological structure—the same open sets, the same limit points, and so forth—and other times different-looking definitions for the neighborhoods of X can result in very different topological structures. When are different-looking topologies equivalent, and when are they different? More generally, given two competing sets of axioms, how are they related? These kinds of questions fascinated Moore, and he spent a lot of his time in pursuit of answers.

Moore's interest remained unchanged throughout a very long academic career. While some mathematicians move from one branch of mathematics to the next over the course of their careers, Moore spent his life studying set-theoretic topology and its logical foundations. In 1916, for example, he published a paper entitled "On the foundations of plane *analysis situs*," and 19 years later he published a paper entitled "A set of axioms for plane *analysis situs*." His masterwork, *Foundations of Point Set Theory*, was originally published in 1932. He revised and reissued it in 1962. Over the course of 46 years, he produced 68 publications, although the large majority of his papers were

## THE MOORE METHOD

When Robert Lee Moore taught a first year graduate-level class in topology, he preferred that his students have little or no previous exposure to the subject. In fact, significant knowledge of topology would sometimes disqualify a student from taking his course; more experienced students would have to take a separate class. Moore also did not want his students to read about topology in their spare time. Instead, the class began with some brief remarks about the axiomatic method. He would write some undefined terms and some definitions on the board, and then he would write a set of axioms and provide a few elementary examples. This was all the preparation that the students received. Throughout the remainder of the course, Moore would write theorems and expect the students to provide proofs. The theorems began at an elementary level but quickly progressed to more advanced work. He would also ask for examples or counterexamples illustrating particular properties of a topological system. He fully expected the students to provide those as well. He did not expect that the class would inevitably rise to the challenge. If some of the students had partial proofs, the class would discuss the ideas that were presented and try to complete them. It was a process, but it was a process that depended on student contributions. If no one had a meaningful contribution to make, class was summarily dismissed. Moore famously described his philosophy with the words "That student is best taught who is told the least."

There were no lectures, no textbooks, and no answers other than what the students themselves provided. As he got to know his students better, Moore followed a definite pecking order: The student he considered weakest was the first one asked to provide a proof. If the student was successful, other methods of proof were sometimes solicited, or the class moved on to the next theorem. If the weakest student was unsuccessful, the next student up from the bottom was asked to continue the proof from the point where the weakest student faltered. Competition—an almost ruthless competition—was encouraged by Moore. Out-of-class collaborations were discouraged. The goal was to spur students to develop their own ideas and insights. Under his guidance, Moore's

produced in the first 23 years. As a young man, he knew what he wanted to study, and he did not deviate from that vision as the decades passed.

students rediscovered the subject (at least as it was understood by Moore) from the beginning. It was the Socratic method applied to 20th-century mathematics.

The Moore method, as his approach became known, has been analyzed and debated for decades. Mathematics departments continue to experiment with it, adopting it, abandoning it, modifying it, and so forth. Some of Moore's students loved it; others were harshly critical of the method. Some hated the competition; some were grateful for the opportunity, but every teaching method has its detractors and its supporters. More fundamentally, Moore's method emphasized mathematics as method rather than as a set of established results and raised the issue of what one learns (or at least what one should learn) when one studies mathematics.

In one famous story about Moore and his method, a prominent mathematician was invited to teach a higher-level summer course at the University of Texas, and after a few classes, he complained to Moore that Moore's graduate students did not seem to know anything. Moore suggested that instead of lecturing his students, he allow them to prove the theorems themselves. The teacher took Moore's suggestion to heart, and in a few weeks—far ahead of schedule—the course was completed. Were Moore's students better educated because they could develop the subject on their own, or was their education deficient because they lacked the background to appreciate the lecture? Reinventing a subject "from scratch" is time consuming, and it leaves little time for sampling other types of mathematics.

Supporters of the method point to the fact that mathematicians educated via Moore's method are generally more active in mathematical research than their more conventionally educated peers when mathematical activity is measured by the number of peer-reviewed papers and books that are published. Critics say Moore's students entered the workforce unaware of broader mathematical trends. As with most disputes, there is some truth on both sides of the question. In retrospect, Moore's critics often displayed a pettiness that did them little credit. It should also be noted that not everyone who could have benefited from the Moore method did so. For a long time, Professor Moore refused to teach African-American students.

Some kinds of mathematics require the researcher to know a large number of techniques and to keep in mind numerous theorems from diverse branches of mathematics. Researchers who specialize in differential equations, for example, are often required to be conversant in topology, analysis, algebra, and numerical analysis—almost every branch of mathematics is used in that discipline. This was not the type of mathematics Moore did, which is not to say that what Moore discovered about topology was simple or easy to accomplish. It was neither, but it is the kind of mathematics in which the problems and solutions are narrowly defined in the sense that neither requires much knowledge of other branches of mathematics. It is the kind of mathematics that is ideally suited to Moore's Socratic method of teaching.

With his research and his style of teaching so perfectly aligned, Moore did not easily accommodate change. It has been mentioned elsewhere that for much of his life he was an unapologetic segregationist. (He was not alone in his views on race. The history of the American Mathematical Society in this regard is not a happy one, and, in fact, Moore, who made little attempt to hide his views about anything, was elected to a term as president of the association.) For many years after he became eligible to retire, Moore continued to teach a full load of classes. Even when the University of Texas cut his pay in half, Moore continued to teach full time. In fact, he taught at the University of Texas until the age of 86, when his employer insisted that he retire. He died shortly before his 92nd birthday.

## Topology in Japan

Before World War II, only a few Japanese mathematicians were active in topology. After the war, Japan became a center for advanced topological research, with special emphasis on set-theoretic topology. It remains an important center of topological thought. The establishment of a Japanese school of topology is due, in large part, to the efforts of one man, Kiiti Morita (1915–95).

Morita's early formal education in mathematics emphasized algebra, and he was very successful as a researcher in that discipline. His name is attached to several important discoveries, especially the concept of Morita equivalence, an idea that is familiar to all researchers specializing in higher algebra. Nevertheless, Kiiti Morita is probably best remembered as a topologist.

Most Western biographies of Kiiti Morita indicate that he turned his attention to topology after becoming a prominent researcher in algebra; this is not quite correct. Although his formal university education emphasized algebra, Morita was attracted to topology while still a student at university. Even as he was enrolled in courses in higher algebra, he was studying topology, but he had few opportunities for formal education in the subject. Undeterred, he undertook a program of self-study. In an act of remarkable intellectual self-confidence, he chose to write his doctoral thesis in topology. He received his doctorate from the University of Osaka in 1950.

*By the 1990s, half of all topologists in Japan had been students of Kiiti Morita or students of his students. His impact as a teacher and researcher on the development of modern topology is profound.* (the Morita family)

Morita contributed to the development of many areas of topology, three of which are described here. First, he helped establish the concept of paracompactness, an extremely important innovation in topology. Paracompactness is a generalization of the concept of compactness, and its discovery is an excellent example of the way that mathematics progresses. Chapter 5 includes a section ("Topological Property 1: Compactness") that describes the concept of compactness, one of the most used and useful concepts in topology. It also includes some examples that indicate why compactness has proven to be such an important concept. Compactness, however, is a very strong condition, and there are many important topological spaces that are not compact. The theorems that apply to compact spaces cannot usually be applied to the study of noncompact spaces. Researchers were faced with a decision. They could put aside the study of compact spaces and study noncompact spaces, or they could attempt to identify those properties that make compact spaces so useful and seek to generalize them. The development of the concept of paracompact spaces is the result of this second strategy. All compact spaces are also paracompact but many spaces that are not compact are paracompact.

To see how paracompactness generalizes the concept of compactness, recall the definition of compactness: Let $X$ be a topological space, and let $S$ be any collection of open sets with the property that every point in $X$ belongs to some element of $S$. The set $S$ is called an open cover of $X$. If it is always true that every open cover $S$ of $X$ contains a finite subcollection that also covers $X$, then (as a matter of definition) $X$ is compact. Let $\hat{S}$ represent the required finite subcollection of $S$. Because $\hat{S}$ contains only finitely many open sets, it is also true that every point in $X$ belongs to only finitely many elements of $\hat{S}$. This observation is the key to understanding paracompactness.

Let $X$ be any topological space—not necessarily compact—and let $S$ be any collection of open sets that cover $X$. If there is always a subset of $S$, which we will again call $\hat{S}$, such that every point in $X$ belongs to at least one but to no more than finitely many elements of $\hat{S}$, then $X$ is called paracompact. The set $\hat{S}$ may be finite or infinite. That is the generalization. Compactness requires that $\hat{S}$ be a finite set; paracompactness requires only that for each $x$ the collection of sets in $\hat{S}$ that contain $x$ must be finite. Experience has shown that this is a very productive idea. All metric spaces, for example, are paracompact. Consequently, any discovery that applies to all paracompact sets must apply to all compact spaces and all metric spaces. Researchers in the fields of differential geometry and algebraic topology have made extensive use of the property of paracompactness. In addition to helping establish the concept, Morita made many useful discoveries about the properties of paracompact spaces.

Morita also made a number of important contributions to the study of metric spaces. The concept of metrizability of a topological space was discussed in chapter 5. A description of Morita's contributions to the study of metric spaces, which are fairly technical, would take us too far afield.

Of special importance to this history is Morita's work in dimension theory. Dimension theory is the subject of chapter 7. Here we only make some general remarks on the context of professor Morita's discoveries about the nature of dimension. One of the great contributions of topologists is the rigorous development of the idea of dimension. As mentioned in chapter 3, discoveries by Dedekind, Cantor, and Peano called into question the idea of dimension. Their work inspired a great deal of research, and during the first half of the 20th century, European mathematicians were able to define several competing notions of dimension. Progress was so rapid that by the 1940s, some mathematicians believed that the subject was largely complete, leaving little room for additional discoveries. Beginning in the 1950s, however, Morita revitalized the subject, extending the concept of dimension to very general topological spaces. He also studied how the differing concepts of dimension are related. Morita proved, for example, that for any metric space, two of the concepts were identical. These are the so-called Čech-Lebesgue dimension and the *large inductive dimension*. (See chapter 7 for a discussion of these concepts of dimension.)

Morita also sought to understand topology from a global point of view by studying the relationships among various classes of topological spaces. To put it another way, he wanted to understand the structure of the field of topology. (See chapter 8 for a discussion of mathematical structure.) Keep in mind that many mathematicians had devoted their time to the study of topology, and from the early years of the 20th century, topological knowledge had grown at an astonishing rate. Many different topological spaces had been described, and by the 1950s, mathematicians interested in topology were faced with an extraordinary array of topological concepts, topological properties, and topological spaces. One way of describing this situation is to say that research had produced a broad array of interesting ideas. Another way to describe the same situation is to say that research had produced an intellectual jumble. To bring order to the array (or the jumble), Morita turned to *category theory*. (See chapter 8 and the afterword for more information on the theory of categories.) The theory of categories had been established only in 1945, and Morita was an early pioneer in the field.

While category theory is far enough removed from topology to be beyond the scope of this book, it is still worthwhile to indicate the general direction of Morita's thinking. As has already been mentioned more than once, early topologists separated the notion of point from the field of geometry. They learned to think in terms of *abstract points*, a term used to convey the idea that points are merely the elements of which a topological space is composed. To them, it did not matter what the term *point* represented or even if it represented anything at all. Category theory goes a step further and suppresses any consideration of points. Instead, the theory of categories concentrates on the nature of certain connections among the different "structures," or in the case considered here, the different classes of spaces. The hope is that the study of these connections, which are called *morphisms*, may reveal something about the relationships that exist among the spaces.

In addition to the many questions about topology that he answered, Kiiti Morita is also remembered for the questions he asked. Most famous are the Morita conjectures, three questions that Morita asked about the properties of certain classes of topological spaces. These questions influenced the development of set-theoretic topology during the last quarter of the 20th century because a number of topologists devoted their energies to the search for answers to Morita's questions.

It is an interesting fact that for decades, even as his results attracted attention in the West, Morita received somewhat less recognition in his home country. He did not leave his native land until the early 1970s, when he briefly taught in the United States. He later lectured occasionally in Europe. Soon, however, he returned to his home in Japan. But if university administrators were somewhat slow to recognize Morita's contributions, university students were not. Morita supervised a number of Ph.D. students who later became prominent in either algebra or topology. They, in their turn, taught others. It is an often-quoted observation that by the middle of the 1990s, half of the topologists in Japan had been taught by Morita, or by Morita's students, or by the students of the students of Kiiti Morita. Like a stone dropped in a still pond, Kiiti Morita's influence expanded across much of Japan. It is still in evidence today.

By all accounts, Kiiti Morita was a quiet, modest, and friendly man who enjoyed the company of his family. In his free time, he enjoyed listening to classical music. Indefatigable as a researcher, he published almost 90 papers during his life, and he was still publishing research into his 70s. It is a remarkable record and a remarkable demonstration of the ability of one individual to make a difference.

# 7

# DIMENSION THEORY

In several places in the preceding chapters, the term *dimension* has been mentioned without trying to be precise about the word's meaning. It is difficult to describe some mathematical ideas without using the word, nor is it just in the field of mathematics that the term is needed. Dimension is a "natural" concept in the sense that if we are to describe accurately how we view the world, we sometimes need to use the idea of dimension. The concept seems to go back at least to the ancient Greeks, and today the word is frequently used in scientific discourse, in science fiction stories, and even in the occasional television commercial. People use the word as if they understand it. Historically, however, it was not easy to develop an unambiguous definition of dimension, and common sense notions of the concept collapse under scrutiny. Mathematicians in several countries struggled to make the definition precise, and when they were finished, they had created three different definitions of dimension. The next steps were to determine the relationships among these definitions and to investigate the logical implications of each definition. The result is the branch of set-theoretic topology called dimension theory.

That the concept of dimension might be subtler than it initially seemed goes back to the writings of Bernhard Bolzano, who attempted to make explicit the concept of dimension. As with many of his other mathematical writings, Bolzano's thoughts on dimension theory failed to influence the development of the subject because he did not elect to publish them, and as with many of his writings, Bolzano was remarkably forward thinking. His early success in dimension theory is even more impressive because he did not know about general topological spaces. His ideas were restricted to subsets of three-dimensional space. He was, therefore, working alone and with few examples of spaces against which to test his ideas. Yet he produced a coherent concept of dimension that is in many ways similar to the small inductive definition described later in this chapter.

Apart from Bolzano, most mathematicians before Cantor believed that the concept of dimension was an "obvious" one. The first publicized indication that the concept of dimension might be more subtle than was widely believed was Cantor's discovery of a one-to-one correspondence between the unit interval and the unit square. (This is described in chapter 3.) Before Cantor's discovery, most, perhaps all, mathematicians believed that what made two-dimensional space two dimensional was that one "needed" two coordinates to identify each point in two-dimensional space. Similarly, three coordinates were used to identify points in three dimensions, and so forth. Cantor, to his own astonishment, discovered that he needed only a single coordinate to identify any point on the unit square, or the unit cube, or even the $n$-dimensional unit hypercube. This proved that the number of coordinates customarily used to identify a point in space did not necessarily indicate the dimensionality of the space in which the point was imbedded. Cantor's correspondence proved the old idea wrong but failed to indicate what the truth might be. For his part, Cantor took some comfort in the realization that his correspondence was discontinuous. Consequently, his correspondence was inadequate as a coordinate system because points that were close together in space could have coordinates that were far apart. Peano's discovery of a continuous space-filling curve further confused the situation and inspired many mathematicians to attempt to develop a mathematically meaningful concept of dimension.

In what follows, it is important to keep in mind that the concept of dimension is an abstraction. Most people would agree that there is some overlap between each mathematical definition of dimension and the physical world, but researchers differ on the nature and extent of the overlap. Each definition of dimension is meant to capture that part of the concept of dimension that

## CONTINUUM THEORY

One subdiscipline of set-theoretic topology that attracted a great deal of interest during the first half of the 20th century was continuum theory. Dedekind's demonstration that the set of real numbers had what he called the *same continuity* as the real line was an important conceptual breakthrough, but it was not the last word on the subject. A continuum, which is defined as a compact, connected subset of a Hausdorff space that contains at least two points, has a number of interesting and surprising properties. The definition of a plane curve, for example, as a continuum that contains no open subsets at the plane permits such strange objects as Sierpiński's gasket. The gasket conforms to the definition of a curve, but it violates most people's common sense notion of what constitutes a curve because every point is a branch point. In fact, over the years, topologists have compiled a long list of strange curves that conform to this more or less standard definition. One might conclude, therefore, that a new definition for the concept of a curve is needed, but alternative definitions have permitted different but equally peculiar objects. Today, part of what it means to learn general topology is to become accustomed to the strange consequences of carefully crafted definitions. The continuum is an example of a simple-sounding concept with exotic implications, but what good is it?

The French mathematician and philosopher Henri Poincairé, one of the most successful mathematicians of the late 19th and early 20th centuries, believed that the continuum was a necessary abstract model for physical space. The physical world is known through observations and experiments, but this way of "knowing" depends on observations that are of finite accuracy; measurements of finite accuracy have their own problems. By way of illustration, Poincairé imagined two points, which he called $A$ and $B$–think of $A$ and $B$ as points on a line–that are

was most important to the mathematicians who developed it. In particular, all of the concepts of dimension that are discussed here depend on the concept of a continuum. Keep in mind that in mathematics, a continuum is a compact connected subset of a Hausdorff space that contains at least two points. (See the sidebar "Continuum Theory.") Continua have a number of counterintuitive properties, too many to recount here, and so rather than try to review the discoveries that were made in continuum theory, just barely distinguishable. In other words, our observations allow us to establish that $A$ is not the same as $B$, but if $A$ were much closer to $B$, they would be too close to tell apart. Consequently, if we choose a point $C$ that is midway between $A$ and $B$, then $C$ will be indistinguishable from both $A$ and $B$. Our observations would, therefore, indicate that $A = C$, $C = B$, and $A \neq B$. (The symbol $\neq$ means "not equal to.") To Poincairé, the idea that $A$ could equal $C$ and $C$ equal $B$ and yet $A$ and $B$ be unequal was intolerable. (Recall Euclid's first axiom: "Things which are equal to the same thing are also equal to one another.") Poincairé's solution was to require that any mathematical model of physical space have the property that one can always distinguish between points, no matter how closely they are positioned, and, in addition, he required that between any pair of points there was always a third point. As a subset of the real line, however, the set of rational numbers has the properties that Poincairé wanted. (Between any two rational numbers, for example, there is always a third.) Yet, even Pythagoras, knew that the set of rational numbers is not sufficient to describe the simplest geometric figures. (The length of the diagonal of the square with sides one unit long, for example, is $\sqrt{2}$, an irrational number.)

Mathematicians responded to these observations by creating the concept of a continuum, a purely mathematical concept. There is no reason to suppose that there exists a physical analogue to a mathematical continuum, nor can any experiment or observation resolve the question of whether space, or time, or space-time forms a continuum in the sense understood by mathematicians. Still, having agreed upon a definition of the concept of continuum, *a definition that satisfied the needs of mathematicians,* topologists spent decades investigating the logical implications of that definition. The result is a remarkable body of theorems that demonstrate just how large a gap separates the discoveries of modern mathematics from common sense notions of continuity. That gap continues to grow.

which are often highly technical, we must rely on examples of continua—especially Richard Dedekind's construction of the real line, which is described in chapter 3—to develop some insight into what continua are. Even so, we would also do well to remember Dedekind's cautious description of the relationship of his mathematical work to physical space. He wrote that the assertion that space is continuous is an axiom. "If space has at all a real existence it is not necessary for it to be continuous; many of its properties would remain the same even were it discontinuous." He did not reject the possibility that space is continuous, he just saw no reason to accept the possibility, either. This is part of the difference between mathematics and the physical world: Whether a subset of a particular topological space forms a continuum depends on the axioms used to define that space; topological spaces do not inherit their continuity from the physical world.

## Inductive Definitions of Dimension

The first general concept of dimension was formulated by the Dutch mathematician and philosopher Luitzen E. J. Brouwer (1881–1966). More philosopher than mathematician, Brouwer devoted his considerable energies to the study of topology for about five years. Not long after he was made a full professor at Amsterdam University in 1912, Brouwer turned his attention to the philosophy of mathematics, and this remained his primary interest for the rest of his life. Some of Brouwer's biographers— and there are several—write that the primary reason he studied topology was to increase his academic profile and draw attention

*Dutch postage stamp honoring Luitzen E. J. Brouwer, one of the most successful topologists of his time, although he preferred philosophy to mathematics* (Netherlands Postal Service)

to his philosophical views. His real interest, the focus of his work before and after the years that he studied topology, was always certain philosophical issues associated with mathematics.

Brouwer's interest in the philosophy of mathematics seems to have been the result of his experiences as a mathematics student at Amsterdam's municipal university. Brouwer initially found mathematics boring—very boring. He characterized mathematical theorems as "Truths, fascinating by their immovability but horrifying by their lifelessness . . ." He began to question the nature of mathematics and what it meant to do mathematics. The logical difficulties that Cantor's investigations had uncovered in his set-theoretic investigations, for example, were to Brouwer the outcome of a misconception about the nature of mathematics. He believed that mathematics consisted of a series of "constructions" that took place in the mind, and he rejected the idea that mathematics was about the logical relationships that existed among words. Brouwer explicitly rejected mathematics as logic. He rejected, for example, the principle of the "excluded third," a principle of logic that states that if $p$ is a mathematical statement, then either $p$ or *not* $p$, the negation of $p$, must be true. In other words, the principal of the excluded middle means that $p$ and *not* $p$ exhaust all possibilities. Brouwer disagreed. His view of mathematics was not widely shared by the mathematicians of his time. Many of his predecessors would have disagreed with him as well. Even Euclid had relied on the principle of the excluded third to complete some of his proofs. It is difficult to do math without it, and Brouwer adopted the principle in his own mathematics when it suited his needs. Despite this apparent inconsistency, Brouwer championed a view of mathematics that he called "intuitionism," a philosophy of mathematics very different from the philosophy that prevails today.

If Brouwer's impact as a philosopher of mathematics was modest, his impact as a mathematician was significant. There are a number of important results in topology associated with his name. In the area of dimension theory, Brouwer was influenced by an observation made by Henri Poincaré. In 1905, Poincaré wrote a philosophically oriented book, *La valeur de la science (The*

*Foundations of Science*), in which he sought to give a plausible—but not a rigorous—definition of dimension. He wanted to answer the question "What does it mean to say that space is of three dimensions?" His answer depends on two concepts: a cut and a continuum. Essentially, Poincaré said that if continuum $A$ can be cut into two disjoint continua by a point, then continuum $A$ is one-dimensional. (Think back, for example, to Dedekind's cuts of the real line.) Having defined what it means to say that a continuum is one-dimensional, Poincaré turned his attention to two-dimensional continua: If continuum $B$ can be cut into two disjoint continua by a one-dimensional continuum, then continuum $B$ is two-dimensional. (Think back, for example, to the Jordan curve theorem, which says that a simple closed planar curve divides the plane into exactly two disjoint regions. One region lies inside the curve, and the other region lies outside.) Having defined what it means to say that a continuum is two-dimensional, he was in a position to define what it means to say that a continuum is three-dimensional: If continuum $C$ can be cut into two disjoint continua by a two-dimensional continuum, then continuum $C$ is three-dimensional. (Think of a sphere. It divides three-dimensional space into two disjoint regions. One lies inside the sphere, and the other lies outside.) The process can be continued to higher dimensions: A continuum is $n$-dimensional if it can be cut into disjoint continua by a continuum of $(n - 1)$ dimensions. Poincaré's ruminations provide an example, if an unfinished one, of an inductive definition. The idea behind an inductive definition is to define a set of terms sequentially, usually beginning with a trivial case, and proceeding up the chain of terms. Each link in the chain is defined in terms of the preceding link. In Poincaré's definition, it moves from dimension one, to two, to three, and so on.

Brouwer used Poincaré's imprecise description of dimension to create an inductive definition of dimension. The problem is that Brouwer's original definition is not quite correct. It fails to correctly apply to certain unusual spaces. These spaces are too technical to describe here, and the first such space was discovered only as Brouwer was completing his paper on dimension theory. Later, Pavel Urysohn found the flaw in Brouwer's work while visiting Brouwer in the Netherlands, and after Urysohn's discovery, Brouwer corrected it. Brouwer's definition has since been further refined, and the corrected version of what Brouwer called *Dimensionsgrad* is now called Ind($X$) because it is an *ind*uctive definition that applies to "large" subsets of a topological space $X$, which explains the capital "I." Ind($X$) is now called the Brouwer-Čech dimension. To be clear: Ind($X$) is a function. Its domain is a class of topological spaces, which are defined later in this section, and its range is the set $\{-1, 0, 1, 2, \ldots\} \cup \{\infty\}$, where the symbol $\infty$ denotes "infinity."

The Bohemian mathematician Eduard Čech (1893–1960) made a number of important contributions to dimension theory as well as other branches of mathematics. (Bohemia was part of the Austro-Hungarian Empire when Čech was born.) He further refined Ind($X$) and identified a large class of topological spaces to which Brouwer's definition applied. A description of Ind($X$), the finished version of Brouwer's *Dimensionsgrad*, follows. It is a nice example of how to define a concept inductively.

The function Ind($X$) applies to a special class of topological spaces called *normal spaces*. The topological space $X$ is normal if for every closed subset $A$ of $X$ and every open subset $V$ containing $A$ there exists an open subset $U$ such that

- $A$ is a subset of $U$, and

- $U$ and the boundary of $U$ belong to $V$.

Recall that the boundary of an open set $U$ consists of those points $x$ with the property that every neighborhood of $x$ contains points that belong to $U$ and points that belong to the complement of $U$. Alternatively, using the concept of limit point, we can say $x$ is a boundary point of $U$ whenever $x$ is a limit point of $U$ and $x$ is a limit point of the complement of $U$. Informally, in a normal space, disjoint closed subsets are always "separate enough" to be surrounded by disjoint open sets. As for Poincairé's description of dimension, the boundary of the open set $U$ is what he used to "cut" each of the spaces that he considered. Instead of looking directly at the dimension of a particular space, Poincairé looked at the dimension of the boundary of the set used to cut it. Brouwer used the same idea.

Here is the updated version of Brouwer's inductive definition of dimension for a normal topological space $X$:

1. To start the definition, let $Ind(X) = -1$ if and only if $X$ is the empty set.

2. Suppose $X$ is not empty. Let $A$ be any closed subset of $X$, and let $V$ be an open subset containing $A$. Because $X$ is a normal space, there is an open set $U$ such that $U$ contains $A$, and $U$ together with its boundary belongs to $V$. We say that $Ind(X) \leq n$ if $Ind$(boundary of $U$) $\leq n-1$. (See the accompanying diagram.)

3. $Ind(X)$ is exactly $n$ if $Ind(X) \leq n$, and it is false that $Ind(X) \leq n - 1$.

4. Finally, $Ind(X)$ is said to be infinite—that is, $Ind(X) = \infty$—if there is no natural number $n$ such that $Ind(X) \leq n$.

Essentially, $Ind(X)$ allows us to move up a chain of spaces, beginning with the empty set. Spaces of lower dimension enable us to classify spaces of higher dimension. If no natural number satisfies condition 2 in the definition, we conclude the space is infinite-dimensional by condition 4.

Compare the definition of $Ind(X)$ with Poincairé's concept. Poincairé had a good idea, but he did not specify to which spaces he meant his idea to apply, nor did he provide an unambiguous method for carrying out the procedure on abstract spaces. $Ind(X)$ is based on Poincairé's idea, but $Ind(X)$ is rigorously defined. It is also a topological property—that is, the large inductive dimension of a topological space is invariant under homeomorphisms. To put it another way: Every topological transformation of a normal space preserves the large inductive dimension of the space.

To return to Brouwer's un-corrected version of large inductive dimension, his *Dimensionsgard*, Brouwer believed that even if his formulation of the concept of dimension was not perfect, it was good enough to qualify him as the founder of dimension theory. This was disputed by the Austrian mathematician Karl Menger (1902–85), who liked to describe Brouwer's work as preparatory, by which he meant that Brouwer's work was a necessary step in preparing the way for the establishment

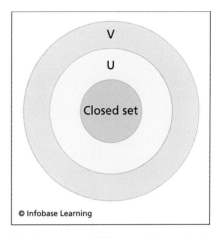

© Infobase Learning

*The boundary of U is one-dimensional. As a matter of definition, therefore, the dimension of the closed set at the center of the diagram is two-dimensional.*

of dimension theory but that it was not enough to qualify as a theory of dimension. Menger developed his own more polished definition of dimension, but his work came after Brouwer's. Not surprisingly, Menger believed that dimension theory started with him, an assertion that Brouwer found very irritating. (Menger's theory of dimension was identical with that of Urysohn.)

The Menger-Urysohn definition of dimension is another example of an inductive definition. A brief biography of Urysohn is given in chapter 5. With respect to Menger, his early work focused on topology, and for a while, he and Brouwer worked at the University of Amsterdam at the same time. Their proximity made it easy for them to argue about who invented dimension theory. Menger eventually left Amsterdam to work at the University of Vienna, and in 1938, he moved to the United States, first working at Notre Dame University and later at the Illinois Institute of Technology, where he remained until his retirement. In later life, Menger's interests shifted to various aspects of geometry, but he is best remembered for his work in topology.

The Menger-Urysohn definition, which is called the small inductive dimension and written ind$(X)$, applies to regular spaces. (Regular spaces were first described in chapter 5, page 91.) As with the Brouwer-Čech large inductive dimension, ind$(X)$ makes use of the boundary of an open set to define dimension. Here is the definition for the small inductive dimension of a topological space $X$:

1. The empty set has dimension –1. In symbols, ind$(\emptyset) = -1$.

2. ind$(X) \leq n$ if for every point $x$ in $X$ and for every open set $V$ containing $x$, there is an open set $U$ containing $x$ such that $V$ contains both $U$ and its boundary and ind(boundary of $U$) $\leq n-1$.

3. ind$(X) = n$ if ind$(X) \leq n$ and it is false that ind$(X) \leq n - 1$.

4. If condition 2 is not satisfied for any natural number, then ind$(X)$ is said to be infinite—that is, ind$(X) = \infty$.

Today, the definition ind$(X)$ is more commonly used than Ind$(X)$. Here are some examples:

Example 7.1. Let $X$ be a finite collection of points in the plane. Draw a circle about each point with the property that all other points in the collection are outside the circle. Then the boundary does not contain any points of $X$—or, to put it another way, the boundary is the empty set. By condition 1, the boundary has small inductive dimension –1, and so the dimension at each point of $X$ is zero. We conclude ind$(X) = 0$.

Example 7.2. Let $X$ be the topological space consisting of the rational numbers. Open sets in $X$ are the intersection of open sets on the real line with the points of $X$. This space has ind$(X) = 0$. To see that this is true, let $x$ be any rational number, and let $V$ be any open set containing $x$. Let $U$ be an open interval centered at $x$ and contained in $V$ and such that $U = \{t: x - r < t <$

$x + r$}, where $r$ is an irrational number. Both $x - r$ and $x + r$ are irrational numbers, so the boundary of $U$ does intersect $X$, the set of rational numbers. That means that in the space of rational numbers, the boundary of $U$ is the empty set. As a matter of definition, ind($\varnothing$) = $-1$, and so we conclude that the dimension of $X$ is zero. (A similar argument shows that the dimension of the topological space consisting of irrational numbers has small inductive dimension zero.)

Example 7.3. Let $X$ represent the circle of radius 1. The topology of the circle consists of all possible unions of open arcs in the circle and all finite intersections of open arcs. (An open arc is an arc that does not contain its endpoints. We have used the open arcs as a basis for the topology on $X$. See chapter 5, page 96, for a description of the idea of a basis.) Let $x$ be any point in $X$ and let $V$ be any open set containing $x$. Let $U$ be an open arc containing $x$ and lying inside $V$, and choose $U$ to be small enough that its boundary also belongs to $V$. (The boundary of $U$ consists of two points, the endpoints of the arc.) As demonstrated in Example 7.1, isolated points have dimension zero. Because ind(boundary of $U$) = 0, it follows that ind($X$) = 1. The circle is one-dimensional.

Example 7.4. Let $X$ represent the plane. Let $x$ be any point in the plane and let $V$ be an open set containing $x$. Draw a small circle around $x$ so that the circle and its contents are inside $V$. Let $U$ be the interior of the circle so that the boundary of $U$ is the circle. By Example 7.3, ind(boundary of $U$) = 1. Therefore, ind($X$) = 2. The plane is two-dimensional.

What is important from the point of view of this history is that mathematicians had to *create* definitions of dimension. They had to give mathematical meaning to the concept. While the idea of dimension may seem "natural," a precise statement of this natural-sounding concept is neither easy nor natural, and a useful definition must conform to certain criteria. It must enable the user to compute the dimension of those spaces that are already known—for example, the dimension of the real line should be 1—or at the very least, if the computation produces results different from those that are generally agreed to be correct, there must be a compelling reason to accept the new results. Both Ind(X) and ind(X), for example, yield reasonable and identical results for common spaces: For both Ind(X) and ind(X), the real line is one-dimensional, the plane is two-dimensional, and as we work our way up into the simpler higher-dimensional spaces, the two functions continue to produce results that (to a mathematician, at least) remain reasonable. Consequently, both Ind(X) and ind(X) agree with our intuition when our intuition is useful, and they produce good results for many spaces that have no readily apparent geometric interpretation.

Of course, none of this explains why ind(X) (or any other definition of the concept of dimension) is important. Why do these definitions matter? One answer is now easy to appreciate: The number ind(X) is preserved under topological transformations. If a space is $n$-dimensional according to the small inductive dimension, its dimensionality will be preserved by all homeomorphisms. Cantor, for example, wanted to show that $n$-dimensional Euclidean space, which is represented by the symbol $E^n$, and $m$-dimensional Euclidean space ($E^m$) are fundamentally different, but he was unable to do so. In modern topological language, he wanted to show that whenever $m$ is not equal to $n$, $E^n$ and $E^m$ are not homeomorphic. If one concentrates on homeomorphisms then (in order to show that $E^n$ and $E^m$ are not homeomorphic), one must demonstrate the nonexistence of a homeomorphism between the two spaces. When the problem is phrased in that way, it is a hard problem to solve, but with a good definition of dimension the proof is easy. Because ind($E^n$) = $n$, any space that is homeomorphic to $E^n$ must also have dimension $n$. Because ind($E^m$) = $m$, $E^m$ cannot be homeomorphic to $E^n$ unless $m$ and $n$ represent the same number. This kind of "dimensional thinking" also disposes of the problems posed by Peano's space-filling curve: Because the small inductive dimension of the unit interval is 1 and the small inductive dimension of the unit square is two, the unit square and the unit interval are topologically different. Consequently, Peano's function cannot be "adjusted" to make it a homeomorphism between the two sets. Similar statements apply to the large inductive dimension.

## A Noninductive Definition of Dimension and More Consequences of Dimension Theory

A very different definition of dimension that was developed at about the same time as ind(X) and Ind(X) is called the Čech-Lebesgue definition of dimension. It was devised by the French mathematician Henri-Léon Lebesgue (1875–1941), one of the most influential mathematicians of the 20th century. Lebesgue began his higher education at a college for teachers. After he graduated, he studied mathematics by himself for about two years. As he studied the work of some of the best mathematicians of his time, Lebesgue realized that he could make a contribution to the field of analysis that was uniquely his. Today, his major contribution is known as the Lebesgue integral, and it is recognized as one of the great achievements of 20th-century mathematics.

As the 19th century drew to a close, mathematicians were struggling to solve problems using techniques that required that the functions under study be continuous, or at least continuous except at a few isolated points. Lebesgue found a way to extend the techniques then in use. His extension agreed with the old techniques when the old techniques were applicable, but it applied in many cases in which the old techniques were inadequate. Lebesgue's work in analysis remains an important part of the education of every serious student of mathematics today.

Lebesgue remained active in mathematics throughout his life, and he did not hesitate to attempt to solve difficult problems in branches of mathematics other than analysis. In particular, he worked in topology, and his contribution to dimension theory is remarkable because it is highly original and very simple. It depends on the notion of an *open cover*. (Open covers were discussed in chapter 5, pages 85–86. The definition is briefly repeated here because it is used in a very different way.) Let $X$ be a normal topological space, and let $\{U, V, W, . . .\}$ be a collection of open subsets of $X$. If every point in $X$ belongs to at least one of the sets in *{U, V, W, . . .}*, then the set *{U, V, W, . . .}* is called an open cover of *X*. There are usually many different open covers of any topological space *X*, and it is usually possible to "refine" every such cover in the following way: Replace *{U, V, W, . . .}* by the open cover *{U', V', W', . . .}* where *U'* is a proper subset of *U*, *V'* is a proper subset of *V*, and so forth. The set *{U', V', W', . . .}* is called a refinement of *{U, V, W, . . .}*.

The order of a cover is the largest number of sets to cover a single point. So, for example, if for a particular cover at least one point belongs to two sets and no point belongs to three sets, then the order of the cover is 2. Here is Lebesgue's definition: If for every open cover of a topological space *X* there is a refinement with order not greater than *n*+1, and *n* is the smallest integer for which this statement is true, then *X* has Čech-Lebesgue dimension *n*. (The Čech-Lebesgue dimension of a topological space is written dim(*X*).)

> Example 7.5. Consider the topological space $\{x: 0 < x < 1\}$. Every open cover of this space that consists of at least two sets has a refinement of order 2. Here, by way of example, is an open cover of $\{x: 0 < x < 1\}$: Let $U = \{x: 0 < x < \frac{2}{3}\}$, and let $V = \{x: \frac{1}{3} < x < 1\}$. The point ½, for example, belongs to both sets. We can refine *U* and *V* in many different ways, but it is impossible to entirely eliminate the overlap and still cover the space. This illustrates the fact that $\{x: 0 < x < 1\}$ has Čech-Lebesgue dimension of 1.

Lebesgue actually stated his theorem only for cubes in $E^n$. It was generalized to a broader class of topological spaces and made more precise by Čech many years later.

Mathematics is often presented with an air of finality—as if the subject appeared in its final state and no alternatives are possible. But in these three definitions of dimension, one can see some of the most astute mathematicians of the 20th century struggling to *create* a concept that is mathematically rigorous and yet does not defy "common sense" notions of what the word dimension means. These mathematicians arrived at three distinct solutions to this problem, and they are distinct, not just in form but in concept.

While they all yield the same dimension for the most common spaces, they give different results when they are applied to less common spaces.

It is tempting to ask which definition is correct, but that assumes that there is a correct definition. There is no single correct definition. Each definition has, however, proved to be useful in the sense that it has helped mathematicians better understand the mathematical "universe" they have created. Cantor's discovery of one-to-one correspondences between sets of different

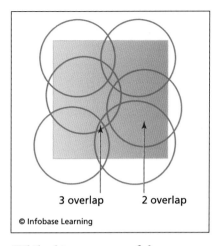

3 overlap    2 overlap

© Infobase Learning

*While this open cover of the square can be further refined, all additional refinements will leave some points of the square simultaneously inside three discs.*

dimensions and Peano's space-filling curve had caused mathematicians to question whether the concept of dimension had any real meaning. The work of Brouwer, Urysohn, Lebesgue, Menger, and Čech resolved those concerns. They showed that dimension, no matter which definition is used, has a reasonable interpretation and is preserved under topological transformations. Once the concept of dimension was rigorously defined, Urysohn showed that the question that had plagued Cantor, the relationship between the cardinality of a set and the dimension of a set, turned out to have an easy answer. Urysohn discovered that the points of any topological space of positive dimension can be placed in one-to-one correspondence with the set of real numbers.

Dimension theory seemed to be a "mature" subject by 1950. The pace of discovery had waned, and while there were still unanswered questions—because there are always some unanswered questions—it did not appear that there was much of importance that remained to be discovered. This was the conclusion of Witold Hurewicz and Henry Wallman, the authors of the 1941 volume *Dimension Theory*, a book that remains an important contribution to the field. They were wrong. Dimension theory was revived in the 1950s, in large measure due to the work of the Japanese school of topologists (see chapter 6). Kiiti Morita proved a number of important theorems in this regard. He extended the classical concept of dimension to new types of spaces. He found new and simpler ways to estimate the dimensions of various spaces, and he discovered some very important facts about the nature of infinite-dimensional spaces. Unfortunately, the discoveries of the Japanese school are quite technical, too technical to summarize here. Because they began their inquiries where the previous generation of topologist left off, their simplest results are often extensions of a highly evolved theory, but the theorems discovered by the Japanese school reinvigorated the subject and demonstrated that there was still a great deal to learn about the concept of dimension. Dimension theory remains an active branch of mathematics today.

## Still Another Concept of Dimension: The Hausdorff Dimension

Felix Hausdorff investigated a definition of dimension different from the three definitions considered already, and as any good definition of dimension should, Hausdorff's definition agrees with the other definitions for the usual simple cases, such as the real line, the plane, and so forth. What distinguishes Hausdorff's concept of dimension is that it need not be a natural number. The *Hausdorff dimension* of a set can be understood as an attempt to characterize the dimensionality of a particular kind of set. They once were called pathological sets; now they are called *fractals*. The person most responsible for popularizing these sets is the Polish mathematician Benoit Mandelbrot (1924–2010). The following briefly examines the nature of these sets in order to develop an appreciation for why it can be worthwhile to assign such sets a fractional dimension.

In describing what a fractal is, Mandelbrot famously used the example of measuring the coast of Britain, and his example is worth repeating. From low Earth orbit, for example, the coast of

*Part of the coast of Britain (and all of the coast of Ireland) as seen from orbit. Because the coastline has the property of self-similarity, the measured length of the coastline depends on the length of the ruler one uses to measure it.* (NASA)

Britain does not look straight. It consists of sections of coastline that are relatively straight connected by sharp changes in direction. We could, therefore, approximate the length of the coast by measuring the lengths of the straight sections and adding them together.

Now suppose that we fly along one of these apparently "straight" sections of coast in an airplane at an altitude of a few miles. At this lower altitude, what looked like a straight section from orbit will not look straight at all. In fact, each section will look very similar to the entire coast as we saw it from orbit. The segments that from orbit appeared to be straight will, from the airplane, appear to consist of many short, straight segments interrupted by sharp changes in direction. The section that we see from the airplane will not, of course, be identical with a section of coast as viewed from orbit, but it will be similar. In fact, if we are shown traces of segments of coastline as viewed from orbit and traces of segments of coastline as viewed from the plane—traces so as to prevent us from identifying the height of the observer from the context of the picture—then we would be unable to distinguish the orbital view from the aerial view. Moreover, if we measure each of these short (aerial) segments and add the segments together, the length of a stretch of coastline when viewed from the plane will be much longer than the length of the same stretch of coastline when viewed from space.

Now isolate a segment of the coast that looked straight when viewed from the airplane. Suppose we walk along that segment and look closely at the boundary that separates land from sea. We would see that the boundary that separates land from sea has the same general shape as the boundary that we saw from space, which is also the shape that we saw from the plane. From our position along the water's edge, the boundary between the land and sea will appear to consist of short straight sections connected by sharp angles, but these straight sections are much shorter than the ones we saw from the plane. If we measure these straight-looking segments and add them together, we obtain a much longer estimate of the coast of Britain than we obtained from viewing it from the plane, which was much longer than it appeared when viewed from space.

Now imagine viewing the line that separates the land from the sea with a magnifying glass. We would see a series of short straight-looking segments joined by sharp curves, the very same properties that we saw when we looked at the coast from orbit. This is a physical approximation of a mathematical property called self-similarity. Each segment of the coast, whether viewed through a magnifying glass or from outer space, exhibits the same jagged property, and the length of the coast depends on how close we are when we measure it.

Mandelbrot did not discover the property of self-similarity. It was well known to early topologists. The Sierpiński gasket—see page 76—is self-similar; it is invariant under changes of scale. No matter how small a piece of the gasket that we magnify and examine, we see essentially the same thing: a mesh with triangular holes of varying sizes oriented according to the same pattern, and the Sierpiński gasket demonstrates another important property of fractals: They are usually generated by following very simple rules. The procedure for creating many common fractals begins with a straight line segment. The segment is symmetrically broken into a collection of segments, and then the rule is applied to each of the individual segments. The result is a larger collection of shorter segments. The procedure is repeated again and again. A simple procedure repeated indefinitely is what causes the figure to be self-similar.

This property of self-similarity stands in sharp contrast to the sort of curves that were studied by earlier generations of mathematicians. They studied curves that are differentiable, and differentiable curves are generally not self-similar. No matter how the curve seems to twist and turn when viewed from afar, at high enough magnification every differentiable curve will be indistinguishable from a straight line. In fact, unless a curve is almost straight over a short enough distance, it will not have a derivative. By contrast, fractals generally fail to have derivatives anywhere because even over very short distances they are never approximately straight.

As has been mentioned at various places in this narrative, topology, because it is so general, allows for a great many sets with unexpected properties, and part of learning topology involves expanding one's notions about what is possible. Hausdorff's dimension, in particular, can be understood as an attempt to characterize the in-between-dimensions property that many fractallike sets have. The mathematics of computing the Hausdorff dimension of a set is beyond the scope of this history, but with respect to the

Sierpiński gasket, the Hausdorff dimension is an irrational number that is approximately 1.585—bigger than the dimensionality of the real line and smaller than the dimensionality of the plane.

When researchers develop mathematical models of the physical world, they use a wide variety of mathematical concepts and techniques. This kind of research is driven by physical considerations, and, consequently, the mathematics is, in a sense, secondary to the science. "Whatever works" is how some researchers describe their choice of mathematics, but that two-word summary obscures a difficulty that all mathematical modelers must address: Every mathematical model of a scientific or engineering phenomenon must be sophisticated enough to reflect those characteristics that are important to the researcher, and the resulting equations must also be simple enough to solve. There is no use in developing a sophisticated set of equations if they are too difficult to solve. It may be surprising, therefore, that despite the increased complexity involved in using fractals and fractallike sets, many researchers have chosen to model nature using surfaces of nonintegral dimension.

A good example of a model that uses surfaces of nonintegral dimension occurs in combustion engineering. Combustion engineering involves controlling the processes by which fuel is burned. Combustion reactions are some of humankind's most basic and important reactions. Most electricity is generated by the burning of fossil fuels, and virtually all transportation is powered by combustion reactions. It is no exaggeration to say that modern life would be impossible without combustion.

There are many types of flame phenomena. When the fuel is a gas, sometimes the air and fuel are mixed together prior to ignition. Natural gas can, for example, be burned in this way. Other times, the fuel is a solid, as is the case for coal and wood, in which case combustion occurs at the fuel-air interface. Other types of combustion phenomena exist, and the properties of each combustion reaction also depend on the air pressure, air temperature, and many other variables. Each type of phenomenon requires its own separate model. Sometimes, researchers have modeled "flame fronts," the surfaces where the fuel is consumed, as smooth two-

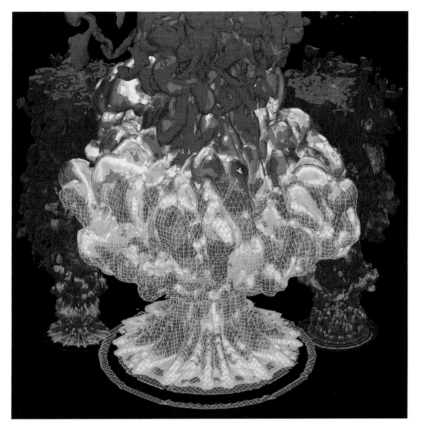

*Computer model of a flame. Flame fronts have complex configurations and a Hausdorff dimension that lies between 2 and 3.* (Paul DesJardin, Department of Mechanical and Aerospace Engineering, University at Buffalo, and Adam Koniak, Center for Computational Research, University at Buffalo)

dimensional surfaces, but combustion engineers have discovered that the situation is more complicated than that. Models that assume the flame front to be "flat" fail to capture certain important properties of the combustion reaction. In a combustion reaction, fuel is often consumed in a thin region that displays what should by now be a familiar mathematical property.

Experiments revealed that under certain circumstances—for example, when the air and fuel are both gaseous and thoroughly mixed prior to ignition—the flame front consists of raised "cells" that resemble individual panels on a quilt. A closer examination reveals that each large cell is composed of numerous raised smaller cells. The cell-like structure of these "premixed" flames appears repeatedly at ever-smaller levels of organization. The flame front exhibits—at least approximately—the same sort of self-similar structure that characterizes the Sierpiński gasket and the coast of Britain.

A word of caution: The Sierpiński gasket is a mathematical construct. It is perfectly self-similar at every scale. Coastlines and flame fronts are not. Just as the self-similar structure that is present in the geography of the British coastline eventually disappears at short enough distances, the nested structure-within-a-structure phenomenon that characterizes mathematical fractals is not present in flames at molecular-scale distances. One should not confuse the model with reality. Still, modeling a flame front as if it had a nonintegral Hausdorff dimension has revealed a great deal of interesting physics, physics that one cannot model if one uses a smooth two-dimensional model of a flame. Researchers have learned more about the velocity of the flame front, the efficiency with which fuel is consumed as the flame moves through the fuel-air mixture, the chemical composition of the products of combustion, and the shape of the flame front by using fractals. And the dimension of the flame front? The answer depends on the pressure, the fuel, the amount of turbulence, and several other factors, but common values of the Hausdorff dimension of a flame front range between 2.2 and 2.4, somewhat more than a surface and substantially less than a volume.

# 8

# TOPOLOGY AND
# THE FOUNDATIONS OF
# MODERN MATHEMATICS

Mathematics is often characterized as consisting of three distinct disciplines, algebra, analysis, and topology, but in addition to being a distinct branch of knowledge, topological ideas have come to permeate much of mathematics. Topology is, for example, the foundation upon which modern analysis is built. In order to appreciate topology, it is necessary to understand why topology has become so central a part of modern mathematics. This chapter describes some of the context in which the field of topology exists.

## Topology and the Language of Mathematics

Before 1860, mathematics was founded largely on geometry—or at least it was founded on geometrical thinking. Part of the reason for this is that geometry was considered more fundamental than other branches of mathematics. As has been mentioned elsewhere in this book, for a long time mathematicians believed that while every mathematical formula could be interpreted as the graph of a function, not every graph could be expressed in terms of a function. These conclusions were not the result of rigorous thinking. They were accepted because they seemed plausible. Unfortunately, plausibility is not the test of mathematical truth, and during the latter years of the 19th century, mathematicians realized that geometrical thinking had led them astray. Ideas of continuity, dimensionality, differentiability, and connectedness

*Georg Cantor and Mrs. Cantor. The former's discoveries placed set theory at the heart of mathematics. As a consequence, set-theoretic topology became an essential part of modern mathematical thought.* (James T. Smith)

could not be understood in geometric terms. To be sure, all of these ideas have their place within geometry in the sense that much of modern geometry cannot be expressed without them, but they are not geometric ideas. Experience showed that geometry was not the right language in which to express these concepts; geometry was not fundamental enough.

A set of points is a far more fundamental notion than a curve, or a surface, or a volume. Every curve, surface, and volume can be described as a set of (geometric) points that satisfies one or more conditions, but there are many sets of geometric points that cannot be interpreted geometrically—at least not in the sense that we can associate a diagram with the point set. The set of points on a coordinate plane with the property that both coordinates are rational numbers is an example of a set that cannot be interpreted as a curve, surface, or volume. This set permeates the plane in the sense that any circle, however small, will contain infinitely many such points, but it can also be shown that almost all of the points on the plane belong to the complement of this set. There are also many sets composed of "abstract points" that are devoid of any connection with geometry—sets of letters, for example, or sets of algorithms. Statements that are true for sets in general will hold for a very broad class of objects, and, in particular, such statements will apply to geometric objects. There are, however, many statements that one can make about geometric objects that have no meaning outside of geometry. As the limitations of geometric thinking became apparent, mathematicians sought an alternative to geometry as the foundation for mathematics. Set theory seemed the natural choice. Perhaps other choices are possible. (See the interview with professor Scott Williams on pages 170–172, for his description of category theory.)

The language of set theory quickly became the language of mathematics. Many of the most prominent mathematicians of the first half of the 20th century devoted at least some of their efforts to the study of sets. Georg Cantor's one-to-one correspondences, although they were an important first step, were only the beginning. Mathematicians soon discovered certain logical errors in Cantor's conception of what constitutes a set. Identifying and correcting these problems and answering other questions, some proposed by Cantor himself, constitute one of the more important strands of 20th-century mathematical thought. As a result of all this activity, set theory grew rapidly.

On the other hand, any theory that purports to apply to everything will lack specificity. In other words, in order for a statement to be generally true, it cannot say too much. This was certainly the case with the theory of sets. Purely set theoretic statements were of little use in geometry or analysis, for example, because they lacked sufficient detail. In order to make their discoveries more useful, mathematicians had to consider narrower classes of sets. This is one way to understand the motivation for creating abstract topological spaces. Mathematicians wanted to retain as much of the generality of set theory as possible, but in order to increase the utility of their theorems, they needed to restrict the classes of sets that they studied. A topological space is still an abstract set of points—that is, it is still a collection of arbitrary objects—but there is additional structure imposed on the set. That "additional structure" is what makes the sets especially useful. A topological space might satisfy only the three standard topological axioms, or it might be a Hausdorff space, or it might be a regular space or a normal space or a metric space. Each additional requirement that one imposes on the set further limits the applicability of any discoveries that might be made about the space. The more properties that a space is required to have, the more detailed the conclusions that one can draw, and all such conclusions will also apply to every other space with the same topological properties. Topology is a balancing act between generality and detail. Both are important.

The language and concepts that topologists created to describe their spaces quickly filtered up to other branches of mathematics. Concepts such as continuity, compactness, and connectedness proved to be extremely useful to mathematicians who studied analysis and geometry. They quickly adopted both the concepts and the language of the topologists. Today, one must use topological notions in discussing analysis and geometry in order to say anything at all.

Topology is sometimes criticized because so many of the results are not practical. There is some truth to this. Even the most enthusiastic topologists often have a difficult time identifying applications for their work. Instead, many topologists respond by saying that a discovery that has no use today may prove to be useful at a later date, and this is true. Historically, this has sometimes happened, but "may prove to be useful" is logically the same as "may not prove to be useful" or "may prove to be useless." Instead, understanding why topology is so crucial to modern mathematics requires that one take a step back. Topology permeates mathematics. Even those who criticize topology for its lack of applications often think topologically; they often use topological results when they create their mathematical models, and they may even use topological language in describing their results, although they may be unaware of any of this. Topology has become the foundation on which much of modern mathematics rests.

## Topology in Analysis

Until the latter half of the 19th century, mathematical analysts were concerned with individual functions. They focused on solving individual equations. This was mathematics as computation. These early analysts discovered new algorithms, new functions, and new mathematical ideas. They also discovered new science. They were so successful that even today many people believe that this is what all mathematicians do, and it is certainly true that many mathematicians continue to learn more about computation. There is even a branch of mathematics, called numerical analysis, dedicated to the study of how to use computers to solve equations. Yet even as computation remains an important part of mathematics, ideas of what it means to "do" mathematics have broadened as mathematicians began to look for broader patterns within mathematics. They wanted, for example, to identify what conditions were necessary and sufficient to ensure the existence of solutions to certain classes of equations. To accomplish this goal, they needed to find a broader, more abstract way of understanding mathematics.

The Italian mathematician Vito Volterra (1860–1940) was one of the earliest mathematicians to attempt to generalize the study of analysis. In 1887, he wrote about real-valued functions defined on sets of curves. In other words, each "point" in the domain of such a function is a curve, and when the function is evaluated at that point (curve), the result is a real number. (By way of example, consider a function $f$ defined on a set of curves of finite length. Let $C$ be one such curve, and let $f(C)$ represent the length of $C$. The function $f$ is a real-valued function defined on the set of curves of finite length.) Volterra's research was important in making the concept of *function* abstract enough that functions could themselves be imagined as points in function spaces, topological spaces in which the points are interpreted as functions. (See also the discussion on research carried out on series of functions by Giulio Ascoli in chapter 3.)

A good example of the close connections between topology and analysis can be found in the work of the French mathematician Maurice Frechet (1878–1973). Frechet had an interesting life. In high school he was taught by the French mathematician Jacques-Salomon Hadamard, who later became a distinguished professor of mathematics but at that time was teaching high school. Student and teacher remained in contact after high school. Hadamard would suggest problems for Frechet to solve, and he would scold him when he made errors. During World War I, Frechet's work as an interpreter—he spoke multiple languages—carried him near the front lines, where the death toll was staggering. He made arrangements to have his complete mathematical works published should he die during the war, but he survived the war and continued his research. He was very creative, and his work influenced many of his successors. With respect to this history, Frechet created what he called "spaces," which he defined as a collection of points together with a set of axioms that defined relations among the points. These spaces can best be understood as abstract models of specific systems of functions. Frechet was a mathematical model builder. He was, for example, the first to define a metric space.

Frechet's definition of a metric is modeled on the ordinary distance function that students learn in junior high school and high school. Recall that if $x$ and $y$ are two points in the plane with coordinates $(x_1, x_2)$ and $(y_1, y_2)$, then the distance between $x$ and $y$, which we will denote by the expression $d(x, y)$, is defined by the expression $d(x, y) = \sqrt{(x_1 - y_1)^2 + (x_2 - y_2)^2}$. (This is just an application of the Pythagorean theorem.) With the properties of this type of "distance function" firmly in mind, Frechet said that if $x$ and $y$ are any two points in a metric space, then "the" distance between $x$ and $y$ is a function that satisfies the following three conditions. (These conditions are also discussed in chapter 5 on page 94. They are reproduced here for ease of reference.)

1. $d(x, y) \geq 0$ and $d(x, y) = 0$ if and only if $x = y$.

2. $d(x, y) = d(y, x)$, which is to say that the distance from $x$ to $y$ equals the distance from $y$ to $x$.

3. $d(x, y) \leq d(x, z) + d(z, y)$.

Frechet's simple definition had an enormous influence on succeeding generations of mathematicians. It inspired a great deal of research in topology to determine conditions on a topological space that are necessary and sufficient to ensure that such a function exists. (See, for example, chapter 5, pages 94–97.) It also enabled mathematicians interested in studying metric spaces of functions to impose topologies on such spaces. A basis for a topology—see page 96 for a definition of *basis*—is formed by the interiors of "spheres." To see how his works, recall that in three-dimensional space, the interior of a sphere of radius $r$ centered at the point $x$ is defined to be the collection of all points that are less than $r$ units from $x$. In symbols, that sentence can be expressed as $\{y: d(x, y) < r\}$, but this definition is completely general. It does not rely on the dimensionality of the space or on our ability to visualize what we are doing. In particular, Frechet's abstract distance function allowed him and his successors to talk about spheres in, for example, certain spaces of functions. For each such space, a topology could be specified using as a basis the interiors of all spheres centered at each point in the space.

The use of an abstract metric also enabled mathematicians to make broad and practical statements about the existence of solutions of certain types of equations. By way of example, we describe a contraction mapping. A contraction mapping is a kind of function. We will call our contraction mapping $f$. The function $f$ has the property that the "mapping," or function, reduces distances in the sense that the distance between two points in the domain—we can call them $x$ and $y$—is always greater than the distance between $f(x)$ and $f(y)$. More formally, suppose that $X$ is a metric space, and let $f$ be a function with the following two properties:

1. The domain of $f$ is $X$, and the range of $f$ is some subset of $X$.

2. If $x$ and $y$ are any two points in $X$, then $d(f(x), f(y)) < md(x, y)$, where $m$ is some positive number smaller than 1, and the same $m$ holds for any choice of $x$ and $y$.

If $f$ satisfies these two properties—that is, if $f$ is a contraction mapping—then under fairly general conditions on the space $X$, we can conclude that there is a unique solution to the equation $x = f(x)$. The proof of this theorem, which is beyond the scope of this history, also indicates how, in principle, the solution can be computed.

Because many practical problems can be reduced to solutions of equations of the form $x = f(x)$, the contraction mapping theorem has practical value. It is not, of course, the final word on the subject. Knowing that a solution exists, or even knowing that an algorithm works "in principle," is not enough to enable a researcher to state what the solution is. Computing the solution to a particular problem may require specialized algorithms, which may require a very different kind of knowledge. The contraction mapping indicates both the advantages and disadvantages in a highly abstract approach to the study of functions.

Frechet's concept of a metric was a tremendous conceptual breakthrough. His work also illustrates the fact that although so much of topology and analysis seems to be highly abstract—and at a certain level it *is* highly abstract—it was created out of very concrete and often very familiar ideas. Abstract formulations of simply stated concrete ideas are often the result of efforts to create idealized models of complex systems. The models are "idealized" in the sense that they retain only the most fundamental properties of the original systems. The vocabulary is chosen to be as inclusive as possible so that research into the model reveals facts about a wide variety of similar systems. Unfortunately, it is often the case that over time the connection between a model and the systems on which it was based is lost, and the interested reader is faced with something that looks as if it were created to be deliberately complicated—deliberately confusing—but the original intention was just the opposite. Often, the model was devised to be simpler and more transparent than any of the systems on which it was based.

Another early attempt to try to formally define a function "space" according to topological principles was carried out early in the 20th century by the Hungarian mathematician Frigyes Riesz (1880–1956). Riesz made important contributions to a number of areas of analysis, and his contributions were widely recognized during his lifetime. He also established the János Bolyai Mathematical Institute

## OTHER KINDS OF TOPOLOGY, OR WHAT HAPPENED TO THE KÖNIGSBERG BRIDGE PROBLEM?

This volume has concentrated on set-theoretic topology, but other mathematical disciplines are also called topology, a fact that leads to some confusion. There is, for example, the field of differential topology, in which mathematicians study those properties of sets that are preserved by diffeomorphisms. A diffeomorphism is a particular type of homeomorphism. In addition to the properties that characterize every homeomorphism, a diffeomorphism (and its inverse) must also be "infinitely differentiable," which means that all of the derivatives of a diffeomorphism and its inverse must exist and be continuous. (So, for example, the derivative of the homeomorphism exists, as does the derivative of the derivative—called the second derivative—and the third derivative of the homeomorphism, and so on.) The requirement that each homeomorphism be a diffeomorphism restricts the subject matter, at least relative to set-theoretic topology, but it also enables the mathematician interested in differential topology to study special classes of sets that have more immediate geometric and physical interpretations than those that commonly arise in set-theoretic topology.

*(continues)*

© Infobase Learning

*Diagram for the Königsberg bridge problem, one of the earliest of all topological problems. The white strips over the "river" represent the bridges of Königsberg.*

## OTHER KINDS OF TOPOLOGY
### (continued)

Another type of topology, called *algebraic topology,* uses methods and ideas that are very different from those employed by set-theoretic topologists. In particular, it makes essential use of ideas from higher algebra. The study of algebraic topology is, therefore, predicated on a thorough study of higher algebra, also called abstract algebra. One of the basic goals of algebraic topology is the classification of surfaces. A rigorous definition of algebraic topology is beyond the scope of this sidebar, but by way of example, consider a sphere, which is the surface of a ball, and a torus, which is the surface of a doughnut. Both are two-dimensional surfaces. Are they also topologically identical? (The answer is no, a fact that is easier to establish with algebraic topology than with set-theoretic topology.)

Throughout this book, spaces were defined to be "topologically identical" provided a homeomorphism existed between them. Sometimes, however, it can be very difficult to determine whether a homeomorphism between the two spaces exists. Certainly, the failure to find a homeomorphism is not a proof that one does not exist. An alternative method of classifying surfaces is to identify the "fundamental group" associated with each surface. (Groups are algebraic objects—their exact definition does not concern us here—and the study of groups constitutes a very

at the University of Hungary in the city of Szeged. (János Bolyai is discussed in chapter 2.) Characteristically, Riesz sought to place his ideas in the most general possible context, and this occurred before the concept of topological space was clearly understood. His "mathematical continuum" is based on four axioms and the term *accumulation point,* which we can take to be a synonym for "limit point."

1. No point in a finite set is an accumulation point of the set.

2. If a point $x$ is an accumulation point of a set $A$ and the set $A$ is contained in the set $B$, then $x$ is an accumulation point of $B$.

3. Let $A_1$ and $A_2$ be disjoint sets, and suppose that $x$ is an accumulation point of their union, then $x$ is an accumulation point of either $A_1$ or $A_2$ or both.

large part of higher algebra.) If the fundamental groups associated with two surfaces are identical, then we can be sure that a homeomorphism exists between the two spaces even if we cannot find it—again, a concept from algebraic topology.

Historically, algebraic topology traces its beginnings to the Königsberg bridge problem, one of the earliest problems of topology. In the early 18th century, the Swiss mathematician and physicist Leonhard Euler (1707–83) considered the following problem: The city of Königsberg (now called Kaliningrad) was divided by two branches of the Pregel River (now the Pregolya River), and the city was united by seven bridges built in the pattern shown in the accompanying diagram on page 151. Euler wanted to know whether it was possible to walk across each bridge in Königsberg exactly once and end at the starting point. He showed that this was impossible.

The Königsberg bridge problem is a topological question. It depends only on the way that the land masses are connected. Distances and shapes, for example, play no role whatsoever. Euler discovered that because each area of land is connected by an odd number of bridges, the walk cannot be completed in the prescribed manner. Today, algebraic topology is a major branch of mathematics with many important applications in mathematics and physics. There is some, but not much, overlap between the methods and ideas of algebraic topology and those of set-theoretic topology.

4. Let $B$ be a set and let $x$ be an accumulation point of $B$. Suppose $y$ is a point different from $x$, then there is a subset $A$ of $B$ such that $x$ is an accumulation point of $A$ but $y$ is not an accumulation point of $A$.

Riesz goes on to describe the concepts of neighborhood, interior point, boundary point, and so forth. This is clearly an attempt to define something conceptually similar to a topological space. Riesz's axioms are quite different from those of Hausdorff, and they are very different from the standard axioms in use today. Riesz's axioms were never widely adopted, but they are not "wrong." They are, instead, a good example of how mathematicians in search of mathematical truth examine the logical consequences of different axioms and different definitions and eventually decide on those that best suit their needs. *Experience* has shown that it is more efficient to establish topological systems using the concept of neighborhoods than of accumulations points, but this was not apparent at the time Riesz did his research. Riesz's efforts represent an interesting attempt to create spaces consisting of generalized points and to express relations among those points in set-theoretic language.

Mathematicians developed a wide variety of "spaces" throughout the 20th century. They are, in fact, still developing them. Each space has certain properties that the investigator finds useful. They are often obtained from previously known spaces by adding an axiom or changing an axiom. We consider only one more. Called a Banach space after the Polish mathematician Stefan Banach (1882–1945), much of the work in creating Banach spaces was actually completed by Frigyes Riesz. Nevertheless, the first careful investigation of Banach spaces was undertaken by Banach and his students, and they are generally given credit for founding the field of functional analysis, a branch of analysis that has become one of the most important disciplines in mathematics. (The contraction mapping theorem described earlier in this section is due to Banach.) Banach's best-known effort in the area of functional analysis is his book called *Theory of Linear Operations.* It has been translated into many languages and can be found in most academic libraries around the world. It is still in print.

Stefan Banach was something of an eccentric. He rarely took the time to write down his ideas. During graduate school, he was often absent from classes. When one of his teachers realized the level at which Banach was thinking, students were assigned to accompany Banach and ask him questions. Banach answered the questions, and the students wrote his responses in the form of mathematics papers that Banach reviewed for accuracy prior to submitting them for publication. After receiving a Ph.D. in mathematics, Banach continued his unusual work habits by "holding court" in a place called the Scottish Café, a nondescript place in L'viv, Poland, now Lvov, Ukraine. Throughout the 1930s, local academics and guests—and some of the finest mathematicians in Europe—spent time at the café.

"Class" began about dinnertime. Participants often worked late into the night discussing mathematics, debating ideas, and sometimes playing chess. For a while they wrote their mathematical ideas on tabletops and napkins, but tabletops and napkins are not "permanent" media—the tabletops, in particular, were wiped clean each night. Some mathematicians complained that important work was being discovered and forgotten at the Scottish Café. Eventually, someone bought a notebook, which was kept at the café. Mathematics problems were proposed and discussed, and if that night's participants thought a problem worthwhile, it was recorded in the notebook along with any solutions if solutions could be found. The *Scottish Notebook* survived World War II, although some of the participants at the Scottish Café did not. Banach, in particular, suffered terrible hardships during the war when the city was under German occupation. He died in 1945, shortly after the Soviet army drove the German army from the city. Today the *Scottish Notebook* has been translated into numerous languages.

*Theory of Linear Operations* contains numerous examples to illustrate Banach's ideas, but the examples also demonstrate why he, Riesz, and their followers created these spaces. Banach spaces were created as abstract models of function spaces. They are, in effect, models of models. The axioms that describe Banach spaces are more complicated than those that define the topological spaces described so far. In particular, Banach spaces have two very different topologies defined on them. One is called the strong topology, which is defined in terms

*Scottish Café. Stefan Banach and his friends gathered here every evening to pursue advanced mathematical research as well as to eat, drink, and play chess.* (Department of Mathematics, University of York)

of a special type of metric called a norm, and the other topology, called the weak topology, defines open sets in an entirely different way. (The description of the weak topology would take us too far afield.) The topologies are not equivalent. Sets that are compact in one topology, for example, may not be compact in the other. Despite their topological complexity, a great many useful results have been obtained by the careful study of Banach spaces, because many problems involving classes of functions can be efficiently rephrased as operations on subsets of Banach spaces.

Mathematicians continue to study the properties of Banach spaces today, and while the language in which they express themselves is sometimes geometric, most Banach spaces have no analogue in the physical world. These "spaces" are models of mathematical thought, and topology is one of the key tools used to understand them. All of this makes descriptions of these spaces hard for the nonspecialist to appreciate. The problem is further exacerbated by the emphasis on abstraction. Large tomes have been written about Banach spaces in which the elements of the spaces are represented by letters, and examples are few or entirely absent. This kind of formal presentation can cause many readers to wonder why anyone would bother studying such objects, but, properly understood, Banach spaces are ideal examples of the power of topological methods in analysis. They enable mathematicians to learn about functions in creative and productive ways. (It is worth noting that Banach's *Theory of Linear Operations* is filled with useful examples of how his ideas can be applied to the analysis of many different spaces.)

## Mathematical Structures and Topology

Much of modern topology and analysis involves the study of spaces. Frechet spaces, Banach spaces, Sobolev spaces, distributional spaces, $L^p$ spaces, Schauder spaces, Orlicz spaces, Hilbert spaces, and many others now occupy the attention of contemporary analysts. Topologists (and others) make frequent reference to compact spaces, regular spaces, normal spaces, paracompact spaces, Baire spaces, Lindelöf spaces, and many others. Each space was created in response to some class of problems. On the face of it, each space looks different from the others, and as a consequence, modern topology and analysis have become littered with spaces. But how different are these spaces from one another, and how are they related?

Beginning in the 1930s, some mathematicians began to express concern that mathematics had become a sort of Tower of Babel. They worried that various branches of mathematics had become incomprehensible to all but a few specialists. Excessive specialization could mean that discoveries in one discipline would have no value outside the discipline—or worse, discoveries in one branch of mathematics would remain within that branch because specialists in other branches of mathematics simply could not understand the discoveries and how they related to their own work. Beginning in the 1930s, some groups of mathematicians began to look for a class of overarching ideas that would enable them to make sense of mathematics "in the large." To understand what these mathematicians tried to do, we begin by considering an example from the life sciences.

Living things can be classified using a series of categories. Each living creature, for example, is either a prokaryote or a eukaryote depending on details of its cell structure. All mammals are eukaryotes, for example, and each mammal belongs either to the subclass Theria, which includes all placental mammals, such as humans and whales and marsupials such as kangaroos and opossums, or it belongs to the subclass Prototheria, which includes the duck-billed platypus and the echidna. Creating an all-encompassing scheme is the job of the taxonomist, and the resulting classificatory system can be represented as a diagram consisting of many nodes and lines connecting the nodes. Each node represents a class of organisms; the lines indicate relationships between classes. Ideally, the diagram could be used to provide a series of tests to enable the user to identify an organism. Given an organism, one begins at the top of the diagram. Is the organism in question a prokaryote or eukaryote? Or perhaps later on: Does the organism meet the criteria for membership in the subclass Theria or the class Protheria? Each answer moves the investigator further along the diagram and further reduces the number of possible species to which the organism might belong. The existence of a detailed taxonomic scheme has made it easier to understand the many ways in which different species are related. It brings order to a complicated system, and it facilitates understanding. The system is not perfect, of course; it is not even complete. It is occasionally revised to take into account new discoveries and new insights, but it has proved to be a very useful idea.

Some mathematicians have sought to create a similar sort of scheme for the "universe" that is modern mathematical thought. People created this universe, but the number of people who understand more than a small corner of it is not large; some claim it is continually decreasing. Central to the idea of understanding mathematics as a whole is the idea of structure, which is the mathematical analogue to taxonomy, and one of the earliest and most influential proponents of this approach is Nicolas Bourbaki.

The first thing to understand about Nicolas Bourbaki is that he does not exist. Bourbaki is the name of a group of prominent French mathematicians. The group was formed in the 1930s over concern that French mathematics had fallen behind the mathematics of other countries, especially Germany. Their initial goal was to write a series of books that would embody classical mathematical knowledge using the most up-to-date mathematical vocabulary and concepts. Their most ambitious project was the 10-volume set called *Eléments de mathématique*. It consisted of the following titles: *Theory of Sets, Algebra, General Topology, Functions of a Real Variable, Topological Vector Spaces, Integration, Lie Groups and Lie Algebras, Commutative Algebra, Spectral Theories,* and *Differential and Analytic Manifolds.*

The members of Nicolas Bourbaki met occasionally to discuss and debate mathematics, philosophy, and the best way to write each of their many volumes. They also published a newsletter in which the debate continued. Perhaps it is not surprising, given the ambitiousness of the task that the membership undertook, that they also tried to "make sense" of mathematics as a whole. The result was the Bourbaki theory of "structure."

The idea was to identify certain properties—they called them mother structures—that are the most basic properties mathematical systems might have. These very basic structures are topological and algebraic. Once the most fundamental properties are identified, the model can be further refined. Bourbaki hoped that their theory of structure would prove to be a mathematician's taxonomy. For example, in the class of topological spaces in which one-point sets are closed, every topological space that is normal (see page 127) is also regular (see page 91), but not every regular space is normal. Consequently, regularity is a more basic property than normality. In the taxonomy of structure, the question of whether a space is regular would, therefore, precede the question of whether the space is normal. (If it is not regular, it cannot be normal.) Or, to take another example, a metric can be used to define a topology, but not every topological space is a metric space. Therefore, the question of whether a set together with a collection of subsets is a topological space would precede the question of whether it is a metric space.

Despite a great deal of fanfare, Nicolas Bourbaki did not complete a theory of mathematical structure, and it remains an open question whether such a theory even exists. Progress, nevertheless, continues to be made. Bourbaki's theory of structure has been replaced by a branch of mathematics begun by the Polish mathematician Samuel Eilenberg (1913–98) and the American mathematician Saunders Mac Lane (1909–2005). Their creation, which is called category theory, traces its origins to a 1945 paper by Eilenberg and Mac Lane. It is a more rigorous set of ideas than Bourbaki's theory of structure, which was always more qualitative than quantitative. Category theory has developed classically, beginning with definitions and axioms and proceeding to a long list of theorems. Category theory is not topology (and so will not be described here), but it can be used to understand some of the relationships that exist among classes of topological spaces. It can be used to bring unity to diversity.

Early in the history of category theory, Kiiti Morita (see page 118) discovered a number of interesting theorems about the overall structure of topology. Morita, a successful algebraist as well as topologist, was uniquely placed to recognize the value of category theory at an early stage of its development and apply the theory to the field of topology. Later, others caught up with him. Today, category theory is a field in its own right as well as a commonly used tool in other mathematical disciplines. As with taxonomy, the theory of categories is not complete, it may not be completable, but it is a step forward in understanding foundational questions in mathematics.

## Conclusion

Set-theoretic topology evolved rapidly throughout the 20th century. The standards of rigor are now very high, and the central questions, posed during topology's first period of development, the period that ended with the onset of World War II, are now fairly well understood. Questions remain, of course, and research continues, but the broad outlines of the set-theoretic landscape seem to be fairly clearly delineated. It took Euclidean geometry 2,000 years to evolve from Euclid's *Elements* to the extremely rigorous axiomatic formulations proposed during the last years of the 19th century—two millennia to answer the questions that the Greeks themselves raised about the fifth postulate and the limits of straight edge and compass geometry. Today, while some questions about Euclidean geometry remain open, all the "big" initial questions have been answered. Set-theoretic topology completed a similar sort of evolution in less than 150 years.

Mathematics has benefited from this effort in three ways. First, topology has provided a new way of thinking about mathematics. Whereas mathematics was once expressed in geometric language, it is now expressed in set-theoretic language. Sets are generally "equipped" with a topological structure, and, consequently, many mathematical systems are at their most fundamental level topological spaces. Topological research has yielded insights into many of these diverse spaces. Second, topology has answered a number of important questions about the nature of continuity and the nature of dimension, and it has revealed previously unsuspected complexity in such simple-sounding mathematical concepts as curves, continua, and connectedness. Finally, set-theoretic topology has been completely incorporated in the field of analysis, for which it serves as the foundation upon which that entire discipline is built.

Despite its importance, set-theoretic topology remains underappreciated. Most mathematics students do not study topology until late in their undergraduate careers, and most other students fail to encounter the subject at all. Popular accounts of topology almost always reduce to the worn-out and misleading expression "rubber sheet geometry," an unfortunate phrase because it contributes little to anyone's understanding of the subject and because without an appreciation for set theory and set-theoretic topology, it is almost impossible to appreciate modern mathematics. Meanwhile, the importance of mathematics in contemporary life continues to grow. It is hoped that this book has increased the reader's awareness of this very important subject and its place in modern mathematics, the "universe that humans built."

# AFTERWORD

## AN INTERVIEW WITH PROFESSOR SCOTT WILLIAMS ON THE NATURE OF TOPOLOGY AND THE GOALS OF TOPOLOGICAL RESEARCH

*Dr. Scott Williams* (Dr. Scott W. Williams)

*Currently a professor of mathematics at the University at Buffalo, Dr. Scott Williams grew up in Baltimore, Maryland, and received a B.S. from Morgan State College (now Morgan State University) and a Ph.D. from Lehigh University. He is a distinguished researcher who has authored numerous papers and lectured at universities and research centers throughout the world. He has also served as consultant to the National Science Foundation, the National Research Foundation, and the Ford Foundation. This interview took place on March 27, 2010.*

**J. T.** Maybe we could start with a definition. In the popular press and even in some encyclopedias, topology is often described as "rubber sheet geometry," which, I think, is not an adequate description. How would you define the subject to which you've devoted so much of your time and energy?

**S. W.** The rubber sheet geometry is just one piece of topology. There is also the attempt to understand the foundations of many areas of mathematics. That is quite a bit different from rubber sheet geometry. As for a definition, I would say that topology is a geometrical exploration of the foundations of mathematics.

**J. T.** OK . . . In preparation for this interview—

**S. W.** Maybe I could even change that.

**J. T.** Please do.

**S. W.** Replace *geometrical* by *spatial*.

**J. T.** And why are you making that change? Why is it significant?

**S. W.** Because some of the origins of topology came by way of geometry. Some came by way of algebra as well. I realize that in addition to my own personal look at mathematics, a fundamental part of mathematics comes from understanding spatial relationships. So when I view things mathematically, I'm also viewing a picture of those things. It may not be a true picture—it may not be possible to draw a true picture—but a picture of things. *Spatial* comes from that.

**J. T.** In preparation for this interview, I read one of your early papers, "The $G_\delta$-topology on Compact Spaces," and I enjoyed it. I thought it was well written. I'm no topologist, but I could understand it, and it treated a problem that I hadn't considered before. But because I'm not a topologist, I couldn't place the paper in context. I couldn't tell how it was related to other questions in topology or analysis, for example. When I realized that I couldn't place the paper in context, it made me wonder. When you study topology, could you be a little more specific about what you hope your research reveals?

**S. W.** We consider many problems. For example, right now my interest has been in infinite-dimensional space, and I'm just trying to understand what various types of properties and constructions mean in these very general settings. I want to understand the consequences of such constructs. You understand that much of mathematics goes on to the exclusion of the "real world." And my claim is that the reason is that the real world is not understood yet.

Many people earlier in mathematics were interested primarily in geometrical structures in two and three dimensions—circles and spheres and cubes or even dodecahedrons, geometrical objects like that. But in the 19th-century, they began to adopt a wider view, a still-incorrect view, of the universe in four dimensions by equating time with one dimension. Now physicists tell us there is a whole lot more going on than that. Maybe it's as many as 11 dimensions, who knows? But in any case, our understanding of that wouldn't be possible if we didn't also have an understanding of the mathematics that goes along with that. So with respect to infinite-dimensional objects and spaces, my work is an exploration of the consequences of certain kinds of constructions and certain kinds of structures.

**J. T.** And what kinds of structures?

**S. W.** I'll *(laughter)* try to give a lay description.

**J. T.** "Structures as they relate to dimension?" is maybe what I should have asked.

**S. W.** The whole question of dimension is interesting in itself. A major surprise came in the 19th century. The surprise was as follows: They thought that for a one-dimensional object, one should be able to describe it as a continuous function of one variable. So, for example, you can think of the unit circle as the set of all points in the plane such that the square of the $x$-coordinate plus the square of the $y$-coordinate equals 1 ($x^2 + y^2 = 1$). That circle is a one-dimensional object because you can find a function from the unit interval into the plane as follows: Each number $t$ goes to the pair ($\cos(2\pi t)$, $\sin(2\pi t)$). And that function of one variable, the $t$s in the unit interval, can be made to describe an entire circle. The shock happened in the 19th century, when mathematicians discovered, to their horror, that things that they thought were two-dimensional could be described in terms of one continuous variable. For example, the unit square and all its innards and the unit disc, which is the circle and all its innards, can be described in terms of a continuous function of one variable. That was a big surprise. So we had to understand what *dimension* meant.

There are many such definitions of *dimension*. Let me try to describe one definition of *dimension*. Let's say that an "empty object" has dimension –1. That's an odd definition, but let's just say that. An object has dimension $n + 1$ provided it has arbitrarily small neighborhoods whose boundaries have dimension $n$. So, for example, the real line has dimension 1 because any point in it has neighborhoods that are open intervals. The boundaries of the open intervals consist of two points, and each point is an object of dimension zero. Those two points have dimension zero because their boundary is empty.

**J. T.**   Yes.

**S. W.**   So I went from dimension –1 to dimension zero to dimension 1. So an object has dimension 2 provided it has arbitrarily small neighborhoods whose boundaries have dimension 1. This is true about the plane. The boundaries of the open discs are circles whose dimension is 1. Why is it 1? Because on the circle each point has an open set, whose neighborhood looks like a little interval. Just as we described the real line as having dimension 1, these circles have dimension 1. And because of that, the disc has dimension 2.

Now we can talk about dimension for every positive integer *n*. I won't talk about fractal dimension, which, in a way, also arose from this exploration—things that have dimensions that are fractions—and we can talk about things that have infinite dimension. And the neighborhoods of these points in the line, plane, three-space, and four-. . . I talked about small discs or they could be spheres making up their boundaries, but I can replace the idea of disc or sphere with the idea of a square or a cube. It's the same thing, discs and spheres or squares and cubes. An infinite-dimensional object might be one that has no finite dimension for any integer *n*, and the boundary of its neighborhood has this same property.

**J. T.** Ah. So the boundary is also infinite-dimensional? Unlike finite dimensional spaces, you don't get a reduction in dimension by passing from the neighborhood of a point to the boundary of the neighborhood?

**S. W.** Yes. There are many definitions of dimension. I only concentrated on one. It's the simplest one I could think of. I think it is even simpler than the so-called Hausdorff dimension. A lot of people look at that as well. I gave you something called the "small inductive dimension."

**J. T.** When you're speaking about dimension and topology in general, this gives you insight into what I'll call "mathematical space." What relationship does mathematical space have to physical space?

**S. W.** With large dimensions like this? Not much at this point. This is what is so interesting about studying the foundations of science. They may have no application that we can see now, but who knows about what happens in five years or 50 years? We don't know.

People study these so-called space-filling curves, which were discovered in the 19th century and proved that even the square and its innards can be described as a continuous function of one variable. That's called a space-filling curve. It turns out that that mathematics is quite useful in digital cameras. You have a way of taking the image and changing it into 0s and 1s and then taking it back out in terms of 0s and 1s. So if you look online for "space-filling curves," you'll probably come across digital cameras somewhere.

**J. T.**  That's true. I remember coming across digital cameras—

**S. W.**  From the 19th century to the 21st century. That thing about space-filling curves—no one had reason to understand any use for it until maybe 1980, about 100 years after it was discovered. So we can't predict when things will happen. We're just making explorations. Then all of sudden the effort is there and present for its use.

**J. T.**  Let me ask you this: Mathematics is often broken up into three branches: topology, algebra, and analysis. How do you see the relationship between topology and the rest of mathematics? Early topological results, for example, were incorporated almost immediately into analysis. Today, how do you see the relationship between your discipline and the rest of mathematics?

**S. W.**  I don't believe that that's a true description of mathematics.

**J. T.**  Oh, how so?

**S. W.**  Mathematics has changed. It had become so—I don't want to say "inbred"—but our new students are learning many different kinds of things. And they are learning how to pull things from many different areas—to study all kinds of things. The segregation of the field is nowhere near as strong as it used to be. When we hire a new faculty member, it's likely to be a topological algebraist who does analysis, if you understand what I'm saying?

**J. T.**  Yes. *(laughter)*

**S. W.**  The area of operations research, which looks to be a very applied area of mathematics, in fact, can use some very deep results from several different fields. One of my very good friends, a fellow named William Massey at Princeton, does operations research, and he often calls me about certain ideas that he thinks may apply to what he does. The old segregation of the field may not be so accurate now.

But on the other hand, what do I do? Well, I've always had a curiosity about all mathematics. At one point my Ph.D. adviser, who to me seemed to know all mathematics but maybe he didn't, said to me, "Too much. You have to focus much more narrowly." And I later understood why he said that, and why I had to focus much more narrowly than just all of mathematics. But I've always had this interest, and you could say this: Any discoveries that one makes may be "by the way." It's often that you go after one thing and "by the way" you discover something completely different. But the idea is that you are continually making this effort and trying to be open enough to discovery to the other thing that might be there "by the way."

**J. T.**  These observations are based on a lot of experience and education, but I'll bet that you didn't know these things when you first started out. So what drew you to the study of topology in the first place?

**S. W.**  I was drawn to mathematics and mathematics research very early in my life. I would guess that my love for that kind of thing probably came sometime during grade school. Probably due to an uncle of mine who used to play numbers games with me, but I don't know for sure. I certainly know that by the time I finished high school, it was clear that I was either going to study music or mathematics, and mathematics was such a very strong draw that I did that and gave up the music. Once I began doing mathematics, I knew that I wanted to be a research mathematician. I didn't know what that meant, but I wanted to explore new things. Even in high school, I invented something which I encountered later in college. It was something called matrices—matrix algebra. I invented a version of matrix algebra in 10th grade. So I've always had this interest towards research.

When I was in college, I had this grand old professor. This was someone who was born in the 19th century—I think 1895, something like that—whose father had been a mathematician. My professor came from the Ukraine, and the father, of course, was also from the Ukraine, and my professor had escaped that area of the world when the communists came in. He was what you used to call a "white Russian" at a Black school, and his interest was algebra.

**J. T.** What was his name?

**S. W.** His name was Vladimir Bohun-Chudyniv, and Bohun-Chudyniv was interested in nonassociative systems in algebra. These were things like loops and quasi groups—all kinds of stuff like that. When I went to graduate school, I thought I'd be doing that. I'd already coauthored a paper with him on those things, and in my senior year in college, all I did was work in—well, I did something in topology but a lot of algebra. So I thought this was what I was going to study, but when I got to graduate school, I found no faculty member the least bit interested in that stuff. And—I hate to bring politics in here but—I had serious questions about racism on the part of algebraists on the faculty there. Maybe it wasn't true, but as a young person, I had this concern. Well, in the meantime I had an absolutely beautiful course in topology. I first talked with an algebraist about working with him, and then the next semester I said, "No, I don't want to work with you. I want to work with a topologist." And that became my area. I studied algebra. I could have written a thesis in algebra, too. I did a lot of algebra in graduate school, but topology was the one I decided to work in.

I think it was an excellent choice for me because of what I would call spatial imagination. Let's say my wife wants me to make some shelves. I can picture the shelves in my mind . . . where they go . . . I can turn it around in two or three dimensions. Understand all there is about those shelves even before I put it on paper, and this is that spatial orientation towards it. I'm very much moved by what I can see with my eyes and then what I can do with it with my imagination. My guess is that this made me a reasonable topologist, whereas an algebraist, as I saw it, involved different kinds of combinatorial abilities, which I loved, but I was nowhere near good enough for that stuff. I didn't know it at the time, but now that I've encountered various colleagues, I see that I was woefully lacking in that ability. *(laughter)*

**J. T.**   That brings me to my next question. When you spoke about building things and imagining their orientations in space . . . what do you see as the relationship between topology and geometry?

**S. W.**   Topology is such a big field, and parts of it are clearly definitely related to geometry but only in the sense that they both require some kind of spatial interrelation and understanding. That is probably the biggest relation between them. When you think about rubber sheet geometry, which would mean homology and homotopy, they clearly have a much stronger relationship with geometry than the topology that I do. The topology that I do is so very close to logic and set theory—I mean, I've written papers that are only set theory. I've gone to set theory conferences, whereas the rubber sheet geometry guys would never do that kind of stuff. They wouldn't be interested in it. My point is that topology is wide enough that I'm having difficulty including all pieces of it in this discussion. Some of it is quite close to geometry—not what I do.

**J. T.**   Concentrating on the type of topology that you do, what do you see as the future for your subject? You mention doing research on infinite-dimensional spaces. Where do you see topology evolving?

**S. W.**   Well, I want to say something about one definition of topology, one that I intentionally did not give, and it's called "the study of abstract geometrical objects and the relations between them or how they may be changed." That last bit is talking about continuous functions or about functions in general. I think that in some areas of topology, functions have not been looked at carefully enough. There has been more interest in the objects that sit there than in the functions between them. When I think of that definition, I think of what I would call one of the greatest innovations of modern mathematics, known as category theory. When I think about category theory, whose fundamental object is actually the map, the relationship, I recognize that in topology as I see it, there hasn't been enough concern with this relationship. It's been more [concern] with the start of the object and the end of the object and not the process of changing it.

**J. T.** That's fascinating. . . . There is one more question that I would like to ask you. There is a certain educational uniqueness to topology. Students encounter analysis in the form of calculus early on—maybe even in high school. And they encounter higher algebra in the form of linear algebra at an early stage of their education—again, maybe even in high school. But topology has a smaller audience. Most students don't encounter it at all unless they're majoring in mathematics, and then they encounter it for the first time later in their undergraduate education—maybe in their third or fourth year of college. Why do you suppose that is? And can you suggest a way that students might learn more about topology earlier in their education?

**S. W.** I wish I had a book I could suggest. The book I'd want to suggest doesn't even introduce students to topology. It introduces students to category theory, which is even worse. *(laughter)* It's a fantastic book. I've been recommending it to all high school students as well as all college students as well as all people interested in mathematics who are 50 years old. The book is called *Conceptual Mathematics*, and the authors are F. William Lawvere and Stephen Schanuel. Now this is not a book about topology, but it is a book about understanding foundations and things in mathematics. Many students from many places in mathematics are now beginning to come to us knowing category theory. This is brand new.

**J. T.** I'm surprised to hear that they arrive knowing category theory.

**S. W.** Many of them do. I teach a topology class. I always have some undergraduates and some graduate students, and I would say roughly 75 percent of the students have encountered category theory before they met me. This is an enormous change. I think that it is the next step. The foundation of mathematics was born with Cantor and set theory, and I think that that has moved onto category theory as forming the foundations of mathematics. So things are done in terms of categories. In my topology class, I talk about the category of metric spaces and distance decreasing maps as opposed to the category of topological spaces and continuous maps. The first one is the category of geometry. The geometrical things that we study there, they are in a topological medium, so we look at them in that fashion.

In my topology class, I have computer scientists, people with computer science majors, physicists, people in economics, and all kinds of people come to take topology from me. It's a wide range of students—undergraduate and graduate students from all these different areas to learn some topology. I don't have a good elementary book. The most elementary book that I know is by James Munkres. I think he is at MIT, or he used to be in any case. He has a book on, I think, elementary topology. What I tell people is go to your bookstore, look at their math book section, and you'll find a paperback on topology. Try to understand it. Try to read it. Those paperbacks are the books that were textbooks when I was a graduate student.

**J. T.** *(laughter)* Thank you very much, Dr. Williams. I very much appreciate the way that you shared your time and insights.

**S. W.** You're welcome.

# CHRONOLOGY

**ca. 3000 B.C.E.**
Hieroglyphic numerals are in use in Egypt.

**ca. 2500 B.C.E.**
Construction of the Great Pyramid of Khufu takes place.

**ca. 2400 B.C.E.**
An almost complete system of positional notation is in use in Mesopotamia.

**ca. 1650 B.C.E.**
The Egyptian scribe Ahmes copies what is now known as the Ahmes (or Rhind) papyrus from an earlier version of the same document.

**ca. 585 B.C.E.**
Thales of Miletus carries out his research into geometry, marking the beginning of mathematics as a deductive science.

**ca. 540 B.C.E.**
Pythagoras of Samos establishes the Pythagorean school of philosophy.

**ca. 500 B.C.E.**
Rod numerals are in use in China.

**ca. 420 B.C.E.**
Zeno of Elea proposes his philosophical paradoxes.

**ca. 399 B.C.E.**
Socrates dies.

**ca. 360 B.C.E.**
Eudoxus, author of the method of exhaustion, carries out his research into mathematics.

**ca. 350 B.C.E.**
The Greek mathematician Menaechmus writes an important work on conic sections.

**ca. 347 B.C.E.**
Plato dies.

**332 B.C.E.**
Alexandria, Egypt, center of Greek mathematics, is established.

**ca. 300 B.C.E.**
Euclid of Alexandria writes *Elements*, one of the most influential mathematics books of all time.

**ca. 260 B.C.E.**
Aristarchus of Samos discovers a method for computing the ratio of the Earth-Moon distance to the Earth-Sun distance.

**ca. 230 B.C.E.**
Eratosthenes of Cyrene computes the circumference of Earth.

Apollonius of Perga writes *Conics*.

Archimedes of Syracuse writes *The Method, On the Equilibrium of Planes*, and other works.

**206 B.C.E.**
The Han dynasty is established; Chinese mathematics flourishes.

**ca. C.E. 150**
Ptolemy of Alexandria writes *Almagest*, the most influential astronomy text of antiquity.

**ca. C.E. 250**
Diophantus of Alexandria writes *Arithmetica*, an important step forward for algebra.

**ca. 320**
Pappus of Alexandria writes his *Collection*, one of the last influential Greek mathematical treatises.

**415**

The death of the Alexandrian philosopher and mathematician Hypatia marks the end of the Greek mathematical tradition.

**ca. 476**

The astronomer and mathematician Aryabhata is born; Indian mathematics flourishes.

**ca. 630**

The Hindu mathematician and astronomer Brahmagupta writes *Brahma Sphuta Siddhānta*, which contains a description of place-value notation.

**ca. 775**

Scholars in Baghdad begin to translate Hindu and Greek works into Arabic.

**ca. 830**

Mohammed ibn-Mūsā al-Khwārizmī writes *Hisāb al-jabr wa'l muqābala*, a new approach to algebra.

**833**

Al-Ma'mūn, founder of the House of Wisdom in Baghdad, Iraq, dies.

**ca. 840**

The Jainist mathematician Mahavira writes *Ganita Sara Samgraha*, an important mathematical textbook.

**1086**

An intensive survey of the wealth of England is carried out and summarized in the tables and lists of the *Domesday Book*.

**1123**

Omar Khayyám, the author of *Al-jabr w'al muqābala* and the *Rubái-yát*, the last great classical Islamic mathematician, dies.

**ca. 1144**

Bhaskara II writes the *Lilavati* and the *Vija-Ganita*, two of the last great works in the classical Indian mathematical tradition.

**ca. 1202**

Leonardo of Pisa (Fibonacci), author of *Liber abaci*, arrives in Europe.

**1360**

Nicholas Oresme, a French mathematician and Roman Catholic bishop, represents distance as the area beneath a velocity line.

**1471**

The German artist Albrecht Dürer is born.

**1482**

Leonardo da Vinci begins to keep his diaries.

**ca. 1541**

Niccolò Fontana, an Italian mathematician, also known as Tartaglia, discovers a general method for factoring third-degree algebraic equations.

**1543**

Copernicus publishes *De revolutionibus*, marking the start of the Copernican revolution.

**1545**

Girolamo Cardano, an Italian mathematician and physician, publishes *Ars magna*, marking the beginning of modern algebra. Later he publishes *Liber de ludo aleae*, the first book on probability.

**1579**

François Viète, a French mathematician, publishes *Canon mathematicus*, marking the beginning of modern algebraic notation.

**1585**

The Dutch mathematician and engineer Simon Stevin publishes "La disme."

**1609**

Johannes Kepler, author of Kepler's laws of planetary motion, publishes *Astronomia nova*.

Galileo Galilei begins his astronomical observations.

**1621**

The English mathematician and astronomer Thomas Harriot dies. His only work, *Artis analyticae praxis*, is published in 1631.

**ca. 1630**

The French lawyer and mathematician Pierre de Fermat begins a lifetime of mathematical research. He is the first person to claim to have proved "Fermat's last theorem."

**1636**

Gérard (or Girard) Desargues, a French mathematician and engineer, publishes *Traité de la section perspective*, which marks the beginning of projective geometry.

**1637**

René Descartes, a French philosopher and mathematician, publishes *Discours de la méthode*, permanently changing both algebra and geometry.

**1638**

Galileo Galilei publishes *Dialogues Concerning Two New Sciences* while under arrest.

**1640**

Blaise Pascal, a French philosopher, scientist, and mathematician, publishes *Essai sur les coniques*, an extension of the work of Desargues.

**1642**

Blaise Pascal manufactures an early mechanical calculator, the Pascaline.

**1654**

Pierre de Fermat and Blaise Pascal exchange a series of letters about probability, thereby inspiring many mathematicians to study the subject.

**1655**

John Wallis, an English mathematician and clergyman, publishes *Arithmetica infinitorum*, an important work that presages calculus.

**1657**

Christiaan Huygens, a Dutch mathematician, astronomer, and physicist, publishes *De ratiociniis in ludo aleae*, a highly influential text in probability theory.

**1662**

John Graunt, an English businessman and a pioneer in statistics, publishes his research on the London Bills of Mortality.

**1673**

Gottfried Leibniz, a German philosopher and mathematician, constructs a mechanical calculator that can perform addition, subtraction, multiplication, division, and extraction of roots.

**1683**

Seki Kōwa, Japanese mathematician, discovers the theory of determinants.

**1684**

Gottfried Leibniz publishes the first paper on calculus, *Nova methodus pro maximis et minimis.*

**1687**

Isaac Newton, a British mathematician and physicist, publishes *Philosophiae naturalis principia mathematica*, beginning a new era in science.

**1693**

Edmund Halley, a British mathematician and astronomer, undertakes a statistical study of the mortality rate in Breslau, Germany.

**1698**

Thomas Savery, an English engineer and inventor, patents the first steam engine.

**1705**

Jacob Bernoulli, a Swiss mathematician, dies. His major work on probability, *Ars conjectandi*, is published in 1713.

**1712**

The first Newcomen steam engine is installed.

**1718**

Abraham de Moivre, a French mathematician, publishes *The Doctrine of Chances*, the most advanced text of the time on the theory of probability.

**1743**

The Anglo-Irish Anglican bishop and philosopher George Berkeley publishes *The Analyst*, an attack on the new mathematics pioneered by Isaac Newton and Gottfried Leibniz.

The French mathematician and philosopher Jean Le Rond d'Alembert begins work on the *Encyclopédie*, one of the great works of the Enlightenment.

**1748**

Leonhard Euler, a Swiss mathematician, publishes his *Introductio*.

**1749**

The French mathematician and scientist Georges-Louis Leclerc, comte de Buffon publishes the first volume of *Histoire naturelle*.

**1750**

Gabriel Cramer, a Swiss mathematician, publishes "Cramer's rule," a procedure for solving systems of linear equations.

**1760**

Daniel Bernoulli, a Swiss mathematician and scientist, publishes his probabilistic analysis of the risks and benefits of variolation against smallpox.

**1761**

Thomas Bayes, an English theologian and mathematician, dies. His "Essay Towards Solving a Problem in the Doctrine of Chances" is published two years later.

The English scientist Joseph Black proposes the idea of latent heat.

**1769**

James Watt obtains his first steam engine patent.

**1781**

William Herschel, a German-born British musician and astronomer, discovers Uranus.

**1789**

Unrest in France culminates in the French Revolution.

**1793**

The Reign of Terror, a period of brutal, state-sanctioned repression, begins in France.

**1794**

The French mathematician Adrien-Marie Legendre (or Le Gendre) publishes his *Éléments de géométrie*, a text that influences mathematics education for decades.

Antoine-Laurent Lavoisier, a French scientist and discoverer of the law of conservation of mass, is executed by the French government.

**1798**

Benjamin Thompson (Count Rumford), a British physicist, proposes the equivalence of heat and work.

**1799**

Napoléon seizes control of the French government.

Caspar Wessel, a Norwegian mathematician and surveyor, publishes the first geometric representation of the complex numbers.

**1801**

Carl Friedrich Gauss, a German mathematician, publishes *Disquisitiones arithmeticae*.

**1805**

Adrien-Marie Legendre, a French mathematician, publishes *Nouvelles méthodes pour la détermination des orbites des comètes*, which contains the first description of the method of least squares.

**1806**

Jean-Robert Argand, a French bookkeeper, accountant, and mathematician, develops the Argand diagram to represent complex numbers.

**1812**

Pierre-Simon Laplace, a French mathematician, publishes *Théorie analytique des probabilités*, the most influential 19th-century work on the theory of probability.

**1815**

Napoléon suffers final defeat at the battle of Waterloo.

Jean-Victor Poncelet, a French mathematician and the "father of projective geometry," publishes *Traité des propriétés projectives des figures*.

**1824**

The French engineer Sadi Carnot publishes *Réflexions sur la puissance motrice du feu*, wherein he describes the Carnot engine.

Niels Henrik Abel, a Norwegian mathematician, publishes his proof of the impossibility of algebraically solving a general fifth-degree equation.

**1826**

Nikolay Ivanovich Lobachevsky, a Russian mathematician and "the Copernicus of geometry," announces his theory of non-Euclidean geometry.

**1828**

Robert Brown, a Scottish botanist, publishes the first description of Brownian motion in "A Brief Account of Microscopical Observations."

**1830**

Charles Babbage, a British mathematician and inventor, begins work on his analytical engine, the first attempt at a modern computer.

**1832**

János Bolyai, a Hungarian mathematician, publishes *Absolute Science of Space*.

The French mathematician Évariste Galois is killed in a duel.

**1843**

James Prescott Joule publishes his measurement of the mechanical equivalent of heat.

**1846**

The planet Neptune is discovered by the French mathematician Urbain-Jean-Joseph Le Verrier from a mathematical analysis of the orbit of Uranus.

**1847**

Georg Christian von Staudt publishes *Geometrie der Lage*, which shows that projective geometry can be expressed without any concept of length.

**1848**

Bernhard Bolzano, a Czech mathematician and theologian, dies. His study of infinite sets, *Paradoxien des Unendlichen*, is first published in 1851.

**1850**

Rudolph Clausius, a German mathematician and physicist, publishes his first paper on the theory of heat.

**1851**

William Thomson (Lord Kelvin), a British scientist, publishes "On the Dynamical Theory of Heat."

**1854**

George Boole, a British mathematician, publishes *Laws of Thought*. The mathematics contained therein makes possible the later design of computer logic circuits.

The German mathematician Bernhard Riemann gives the historic lecture "On the Hypotheses That Form the Foundations of Geometry." The ideas therein play an integral part in the theory of relativity.

**1855**

John Snow, a British physician, publishes "On the Mode of Communication of Cholera," the first successful epidemiological study of a disease.

**1859**

James Clerk Maxwell, a British physicist, proposes a probabilistic model for the distribution of molecular velocities in a gas.

Charles Darwin, a British biologist, publishes *On the Origin of Species by Means of Natural Selection*.

**1861**

Karl Weierstrass creates a continuous nowhere differentiable function.

**1866**

The Austrian biologist and monk Gregor Mendel publishes his ideas on the theory of heredity in "Versuche über Pflanzenhybriden."

**1872**

The German mathematician Felix Klein announces his Erlanger Programm, an attempt to categorize all geometries with the use of group theory.

Lord Kelvin (William Thomson) develops an early analog computer to predict tides.

Richard Dedekind, a German mathematician, rigorously establishes the connection between real numbers and the real number line.

**1874**

Georg Cantor, a German mathematician, publishes "Über eine Eigenschaft des Inbegriffes aller reelen algebraischen Zahlen," a pioneering paper that shows that all infinite sets are not the same size.

**1890**

The Hollerith tabulator, an important innovation in calculating machines, is installed at the United States Census for use in the 1890 census.

Giuseppe Peano publishes his example of a space-filling curve.

**1894**

Oliver Heaviside describes his operational calculus in his text *Electromagnetic Theory*.

**1895**

Henri Poincaré publishes *Analysis situs*, a landmark paper in the history of topology, in which he introduces a number of ideas that would occupy the attention of mathematicians for generations.

**1898**

Émile Borel begins to develop a theory of measure of abstract sets that takes into account the topology of the sets on which the measure is defined.

**1899**

The German mathematician David Hilbert publishes the definitive axiomatic treatment of Euclidean geometry.

**1900**

David Hilbert announces his list of mathematics problems for the 20th century.

The Russian mathematician Andrey Andreyevich Markov begins his research into the theory of probability.

**1901**

Henri-Léon Lebesgue, a French mathematician, develops his theory of integration.

**1905**

Ernst Zermelo, a German mathematician, undertakes the task of axiomatizing set theory.

Albert Einstein, a German-born American physicist, begins to publish his discoveries in physics.

**1906**

Marian Smoluchowski, a Polish scientist, publishes his insights into Brownian motion.

**1908**

The Hardy-Weinberg law, containing ideas fundamental to population genetics, is published.

**1910**

Bertrand Russell, a British logician and philosopher, and Alfred North Whitehead, a British mathematician and philosopher, publish *Principia mathematica*, an important work on the foundations of mathematics.

**1913**

Luitzen E. J. Brouwer publishes his recursive definition of the concept of dimension.

**1914**

Felix Hausdorff publishes *Grundzüge der Mengenlehre.*

**1915**

Wacław Sierpiński publishes his description of the now-famous curve called the Sierpiński gasket.

**1917**

Vladimir Ilyich Lenin leads a revolution that results in the founding of the Union of Soviet Socialist Republics.

**1918**

World War I ends.

The German mathematician Emmy Noether presents her ideas on the roles of symmetries in physics.

**1920**

Zygmunt Janiszewski, founder of the Polish school of topology, dies.

**1923**

Stefan Banach begins to develop the theory of Banach spaces.

Karl Menger publishes his first paper on dimension theory.

**1924**

Pavel Samuilovich Urysohn dies in a swimming accident at the age of 25 after making several important contributions to topology.

**1928**

Maurice Frechet publishes his *Les espaces abstraits et leur théorie considérée comme introduction à l'analyse générale,* which places topological concepts at the foundation of the field of analysis.

**1929**

Andrey Nikolayevich Kolmogorov, a Russian mathematician, publishes *General Theory of Measure and Probability Theory,* establishing the theory of probability on a firm axiomatic basis for the first time.

**1930**

Ronald Aylmer Fisher, a British geneticist and statistician, publishes *Genetical Theory of Natural Selection*, an important early attempt to express the theory of natural selection in mathematical language.

**1931**

Kurt Gödel, an Austrian-born American mathematician, publishes his incompleteness proof.

The Differential Analyzer, an important development in analog computers, is developed at Massachusetts Institute of Technology.

**1933**

Karl Pearson, a British innovator in statistics, retires from University College, London.

Kazimierz Kuratowski publishes the first volume of *Topologie*, which extends the boundaries of set theoretic topology (still an important text).

**1935**

George Horace Gallup, a U.S. statistician, founds the American Institute of Public Opinion.

**1937**

The British mathematician Alan Turing publishes his insights on the limits of computability.

Topologist and teacher Robert Lee Moore begins serving as president of the American Mathematical Society.

**1939**

World War II begins.

William Edwards Deming joins the United States Census Bureau.

The Nicolas Bourbaki group publishes the first volume of its *Éléments de mathématique*.

Sergey Sobolev elected to the USSR Academy of Sciences after publishing a long series of papers describing a generalization of the concept of function and a generalization of the concept of derivative. His work forms the foundation for a new branch of analysis.

**1941**

Witold Hurewicz and Henry Wallman publish their classic text *Dimension Theory*.

**1945**

Samuel Eilenberg and Saunders Mac Lane found the discipline of category theory.

**1946**

The Electronic Numerical Integrator and Calculator (ENIAC) computer begins operation at the University of Pennsylvania.

**1948**

While working at Bell Telephone Labs in the United States, Claude Shannon publishes "A Mathematical Theory of Communication," marking the beginning of the Information Age.

**1951**

The Universal Automatic Computer (UNIVAC I) is installed at U.S. Bureau of the Census.

**1954**

FORmula TRANslator (FORTRAN), one of the first high-level computer languages, is introduced.

**1956**

The American Walter Shewhart, an innovator in the field of quality control, retires from Bell Telephone Laboratories.

**1957**

Olga Oleinik publishes "Discontinuous Solutions to Nonlinear Differential Equations," a milestone in mathematical physics.

**1965**

Andrey Nikolayevich Kolmogorov establishes the branch of mathematics now known as Kolmogorov complexity.

**1972**

Amid much fanfare, the French mathematician and philosopher René Thom establishes a new field of mathematics called catastrophe theory.

**1973**

The C computer language, developed at Bell Laboratories, is essentially completed.

**1975**

The French geophysicist Jean Morlet helps develop a new kind of analysis based on what he calls "wavelets."

**1980**

Kiiti Morita, the founder of the Japanese school of topology, publishes a paper that further extends the concept of dimension to general topological spaces.

**1982**

Benoît Mandelbrot publishes his highly influential *The Fractal Geometry of Nature.*

**1989**

The Belgian mathematician Ingrid Daubechies develops what has become the mathematical foundation for today's wavelet research.

**1995**

The British mathematician Andrew Wiles publishes the first proof of Fermat's last theorem.

JAVA computer language is introduced commercially by Sun Microsystems.

**1997**

René Thom declares the mathematical field of catastrophe theory "dead."

**2002**

*Experimental Mathematics* celebrates its 10th anniversary. It is a refereed journal dedicated to the experimental aspects of mathematical research.

Manindra Agrawal, Neeraj Kayal, and Nitin Saxena create a brief, elegant algorithm to test whether a number is prime, thereby solving an important centuries-old problem.

## 2003

Grigory Perelman produces the first complete proof of the Poincaré conjecture, a statement about some of the most fundamental properties of three-dimensional shapes.

## 2007

The international financial system, heavily dependent on so-called sophisticated mathematical models, finds itself on the edge of collapse, calling into question the value of the mathematical models.

## 2008

Henri Cartan, one of the founding members of the Nicolas Bourbaki group, dies at the age of 104.

# GLOSSARY

**algebraic topology** a branch of **topology** that uses techniques from higher algebra to analyze topological spaces

**analysis** that branch of mathematics that grew out of calculus and includes differential equations and **functional analysis**

**analysis situs** an expression coined by Gottfried Leibniz for a new conception of geometry. Leibniz's idea was refined and evolved into **topology.**

**axiom** a statement that forms a basis for a deductive argument

**basis** a collection of subsets of a set $X$ that is used to specify the topology on $X$

**boundary** Let $A$ be a set in a **topological space** $X$. The boundary of $A$ consists of those points that are limit points of $A$ and also limit points of the complement of $A$.

**bounded set** In a metric space, a set $A$ is bounded if it is a subset of a set of the form $\{x: d(x,0) \leq n\}$ for some $n$.

**category theory** a formal theory that is concerned with identifying the relationships among various parts of mathematics

**closed set** Let $X$ be a **topological space** and let $A$ be a subset of $X$. The set $A$ is closed provided the complement of $A$ is open.

**compact set** a subset $A$ of a **topological space** with the property that every open cover of $A$ has a finite subcollection that also covers $A$

**complement** Let $X$ be a set, and let $A$ be a subset of $X$. The complement of $A$ (relative to $X$) is the set of points belonging to $X$ that are not in $A$.

**connected set**   Let $X$ be a **topological space,** and let $A$ be a subset of $X$. The set $A$ is connected provided $A$ cannot be represented as the union of two open disjoint nonempty sets.

**continuous function**   Let $X$ and $Y$ be two **topological spaces.** A function $f$ with domain $X$ and range $Y$ is continuous if for every open set $V$ in $Y$ there is an open set $U$ in $X$ such that $f(x)$ belongs to $V$ whenever $x$ belongs to $U$.

**continuum**   a compact connected set that contains at least two points

**countable set**   a set that can be placed in one-to-one correspondence with the natural numbers

**cover**   Given a **topological space** $X$ and a subset $A$ of $X$, a cover of $A$ is a collection of sets, usually taken to be open, with the property that the union of the cover contains $A$.

**Dedekind cut**   a procedure conceived by Richard Dedekind in which a one-to-one correspondence is formed between points on the line and the set of real numbers

**de Morgan's laws**   set-theoretic statements showing the relations that exist between the operations of union, intersection, and complementation

**derivative**   the slope of a line tangent to a curve or the **function** that provides such information

**differentiable**   A **function** is differentiable if at each point interior to its domain, it has a derivative.

**dim(X)**   notation for the noninductive Čech-Lebesgue dimension

**dimension**   the property of spatial extension

**dimension theory**   a subdiscipline of set-theoretic topology that provides axiomatic models for the concept of dimension

**disjoint**   Two sets are disjoint if their intersection is empty.

**empty set**   a set with no elements

**equicontinuity**   A set of **functions** with a common domain and range is equicontinuous if, given a positive number, usually represented by the letter $\varepsilon$, there exists another positive number, usually represented by the letter $\delta$, such that whenever the distance between $x_1$ and $x_2$ is less than $\delta$, the distance between $f(x_1)$ and $f(x_2)$ is less than $\varepsilon$ and the same $\delta$ and $\varepsilon$ will work for every pair of points $x_1$ and $x_2$ in the domain and for every function $f$ in the set.

**Euclidean space**   For $n = 1,2,3, \ldots$ $n$-dimensional Euclidean space is the set of all ordered $n$-tuples of real numbers together with the following $n$-dimensional metric: the distance between $(x_1, x_2, \ldots, x_n)$ and $(y_1, y_2, \ldots, y_n)$ is defined as $\sqrt{(x_1 - y_1)^2 + (x_2 - y_2)^2 + \ldots + (x_n - y_n)^2}$.

**Euclidean transformation**   a translation, rotation, or reflection of Euclidean space or of a figure in Euclidean space. Euclidean transformations are characterized by the fact that they do not change distances and angular measurements.

**$F_\sigma$ set**   A generalization of the concept of a closed set, a $F_\sigma$-set is the union of a countable collection of closed sets.

**function**   a collection of ordered pairs with the property that the first element of each ordered pair appears exactly once in the collection

**functional analysis**   the study of **topological spaces** in which the points are taken to represent functions

**$G_\delta$ set**   A generalization of the concept of open set, a $G_\delta$-set can be written as a countable intersection of open sets.

**Hausdorff dimension**   a concept of **dimension** in which some sets have fractional dimension

**Hausdorff space**   a **topological space** with the property that for any two points $x$ and $y$ in $X$, there are open disjoint sets $U$ and $V$ such that $U$ contains $x$ and $V$ contains $y$

**homeomorphism**   a **function** that is a one-to-one continuous correspondence with the additional property that the inverse function is also continuous. Homeomorophisms are topological transformations.

**ind(X)**   notation for the small **inductive dimension**

**Ind(X)**   notation for the large **inductive dimension**

**infinite set**   any set that can be placed in one-to-one correspondence with a proper subset of itself

**interior point**   Let $X$ be a **topological space,** and let $A$ be a subset of $X$. Let $x$ be a point in $A$, then $x$ is an interior point of $A$ if there is a neighborhood containing $x$ that belongs to $A$.

**intersection**   Let $S$ be any collection of sets. The intersection of the elements of $S$ is a set $Z$ that contains only those points common to all elements in $S$.

**inverse function**   Let $f$ denote a **function.** The inverse function of $f$, if it exists, is obtained by interchanging the positions of the elements in each ordered pair comprising $f$. The inverse function is often denoted by $f^{-1}$.

**large inductive dimension**   a method of defining the dimension of a normal space $X$ in terms of the dimension of the boundary of certain open sets contained within $X$

**limit point**   A point $x$ is a limit point of a set $A$ in a **topological space** $X$ provided every neighborhood of $x$ contains points of $A$ different from $x$. The point $x$ may or may not belong to $A$.

**metric**   a function defined on pairs of points in a set $X$ such that the value of the metric is interpreted as the distance between the points

**metric space**   a **topological space** in which the **topology** is defined by a metric

**neighborhood**   an **open set**

**normal space**   Let $X$ be a **Hausdorff space.** The space $X$ is normal if for every closed set $A$ in $X$ and every open set $V$ containing $A$, there is an open set $U$ such that $U$ and its boundary are a subset of $V$.

**nowhere differentiable function**   a **function** with the property that at no point does a derivative exist

**one-to-one correspondence**  Let $A$ and $B$ be sets. A one-to-one correspondence between $A$ and $B$ is a collection of ordered pairs $(a, b)$ such that $a$ belongs to $A$, $b$ belongs to $B$, and every element of $A$ and every element of $B$ occurs exactly once among the set of all ordered pairs.

**open cover**  Let $A$ be a subset of a **topological space** $X$. An open cover of $A$ is any collection of open sets with the property that $A$ is a subset of the union of the sets in the collection.

**open set**  a subset of a **topological space** with the property that every point in the set is an interior point of the set

**partition**  Let $X$ be a set. A partition of $X$ consists of a collection of subsets of $X$ with the property that each point in $X$ belongs to exactly one set in the collection.

**point**  an element of a set

**real-valued function**  any **function** with the property that the range is a subset of the real numbers

**regular space**  Let $x$ be any point in a **Hausdorff space** $X$. Let $V$ be an open set containing $x$. The space $X$ is regular provided there always exists an open set $U$ containing $x$ such that $U$ and its boundary belong to $V$.

**set-theoretic topology**  the study of the properties of sets of abstract points equipped with a topology

**space-filling curve**  a **function,** the domain of which is a subset of the real numbers and the range of which contains an open subset of the plane

**transformation**  a **function.** A topological transformation is a function that preserves topological properties—also called a **homeomorphism.** Alternatively, topology *is* the study of those properties preserved by topological transformations.

**topological space**  A set $X$ of points together with a collection of subsets of $X$. The collection of subsets, called the **topology** of $X$, satisfies three properties: (1) both $X$ and the empty set belong to the collection, (2) the intersection of any finite subcollection of elements in the topology belongs to the topology, and (3) the union of any subcollection of elements in the topology belongs to the topology.

**topology** 1. the collection of all open sets in a **topological space** 2. the mathematical discipline dedicated to the study of those properties of a topological space that remain unchanged under the set of all homeomorphisms

**union** Let $S$ be a collection of sets. The union of the collection is the set $Z$ consisting of every **point** that belongs to at least one element of $S$.

**unit cube** every **point** in three-dimensional Euclidean space belonging to this set: $\{(x, y, z): 0 \le x \le 1; 0 \le y \le 1; 0 \le z \le 1\}$

**unit interval** every **point** on the real line belonging to this set: $\{x: 0 \le x \le 1\}$

**unit square** every **point** in the plane belonging to this set: $\{(x, y): 0 \le x \le 1; 0 \le y \le 1\}$

# FURTHER RESOURCES

## MODERN WORKS

Not much has been written about topology for a general audience. Most of the classic histories of mathematics stop at the beginning of the 20th century. Most books that have been written about topology assume that the reader has already acquired a strong mathematical background, and, consequently, they can be very challenging for those who have not yet acquired that background. The following is a list of books and articles written for a general audience as well as books and articles that are useful for acquiring a stronger background.

Aczel, Amir D. *The Mystery of the Aleph: Mathematics, the Kaballah, and the Human Mind.* New York: Four Walls Eight Windows, 2000. This is the story of how mathematicians developed the concept of infinity. It also covers the religious motivations of Georg Cantor, the founder of modern set theory.

Boaz, Ralph P. *A Primer of Real Functions.* Washington, D.C.: Mathematical Association of America, 1981. This small book was written as an introduction to analysis, but the first chapter, which is about one-third of the book, constitutes an excellent introduction to set-theoretic topology as it is used in analysis. While the author assumes that the reader has completed a course in calculus, no calculus is needed for the first chapter. Highly recommended.

Flegg, H. Graham. *From Geometry to Topology.* Mineola, N.Y.: Dover Publications, 2001. This brief book is divided into 17 chapters. The first three move the reader from geometry to topology. Chapters 4 through 11 are an introduction to some important concepts in algebraic topology, and the remaining five chapters provide a brief but respectable introduction to set-theoretic topology.

Gardiner, Martin. *The Colossal Book of Mathematics.* New York: Norton, 2001. Martin Gardner had a gift for seeing things mathematically. This "colossal" book contains sections on topology, logic, geometry, and more.

Halmos, Paul R. *Naïve Set Theory*. New York: D. Van Nostrand, 1960. Reprinted many times since it was first published in 1960, this book remains a good introduction to the theory of sets written by a mathematician who loved the subject.

Lawvere, F. William, and Stephen Hoel Schanuel. *Conceptual Mathematics: A First Introduction to Categories*. New York: Cambridge University Press, 2009. This is the book that Professor Scott Williams recommended so highly in the afterword.

Manheim, Jerome H. *The Genesis of Point Set Topology*. New York: Macmillan, 1964. The bad news is that this book is hard to find. The good news is that it provides a well-written description of the very earliest ideas in the history of set-theoretic topology, also known as point-set topology.

Munkres, James R. *Topology, a First Course*. Englewood Cliffs, N.J.: Prentice-Hall, 1975. This is one of the books that Professor Scott Williams mentioned in the afterword. Written for college seniors and first-year graduate students, it is not usually described as "elementary."

Tabak, John. *Numbers*. Rev. ed. New York: Facts On File, 2011. Some of Cantor's most famous one-to-one correspondences are demonstrated in this volume of The History of Mathematics set as well as a detailed description of Russell's paradox, the first logical contradiction to be detected in Georg Cantor's formulation of set theory.

Wilder, Raymond L. "The Axiomatic Method." In *The World of Mathematics*, Vol. 3, edited by James R. Newman. New York: Dover Publications, 1956. This is a fairly technical account of the axiomatic method, but it is well-worth reading.

## ORIGINAL WORKS

Reading a discoverer's own description can sometimes deepen our appreciation of an important mathematical discovery. Often, this is not possible, because the original description is too technical, but sometimes the idea does not require much technical background to be appreciated. Other times, a discoverer will write a nontechnical account of his or her work for a general readership. Here are some classics.

Bolzano, Bernhard. *Paradoxes of the Infinite*. London: Routledge & Paul, 1950. Bolzano is not an easy author to read, but this book is worth the effort because it was so far ahead of its time. While mathematicians today will disagree with some of Bolzano's conclusions, many of his ideas are correct, and he anticipated the work of other better-known mathematicians by decades. Bolzano, who was a priest, also inserts a lengthy section about metaphysics. The text is accompanied by commentary by Hans Hahn, one of whose own articles is referenced below. You can find this book in any academic library or through interlibrary loan.

Dedekind, Richard. "Irrational Numbers." In *The World of Mathematics*, Vol. 1, edited by James R. Newman. New York: Dover Publications, 1956. This is an excerpt from Dedekind's famous work *Continuity and Irrational Numbers*, in which he demonstrates that the real number system has "the same continuity as the straight line." Highly recommended.

Euclid of Alexandria. *Elements*. Translated by Sir Thomas L. Heath. Great Books of the Western World, Vol. 11. Chicago: Encyclopedia Britannica, 1952. One of the most influential books in history, Euclid's work remains an excellent introduction to the axiomatic method.

Euler, Leonhard. "The Seven Bridges of Königsberg." In *The World of Mathematics*, Vol. 1, edited by James R. Newman. New York: Dover Publications, 1956. This is one of the earliest of all articles describing a topological problem, and it is written by one of the most creative and prolific mathematicians in history.

Galileo Galilei. *Two New Sciences*. Translated by Henry Crew and Alfonso deSalvio. Mineola, N.Y.: Dover Publications, 1954. Still in print, this scientific classic is written in the form of a dialogue among three fictitious characters. The dialogue takes place over four days. The first day contains Galileo's observations on infinite sets.

Hahn, Hans. "The Crisis in Intuition." In *The World of Mathematics*, Vol. 3, edited by James R. Newman. New York: Dover Publications, 1956. Hans Hahn was a successful mathematician with a gift for writing popular accounts of mathematics. In this article, he describes how to obtain a nowhere differentiable curve, Peano's space-filling curve, Sierpiński's gasket, and several other counterintuitive mathematical objects. He also places these objects in a historical context so that the reader can better understand why they are important.

Poincaré, Henri. *The Foundations of Science: Science and Hypothesis, The Value of science, Science and Method.* Translated by George Bruce Halsted. New York: The Science Press, 1913. This book, written for a general audience by one of the great mathematicians in the history of the subject, contains Poincaré's writings about the goals of topology, which he calls *analysis situs.*

———— *The Value of Science.* Translated by George Bruce Halsted. New York: Dover, 1958. First appearing in English in 1907, this once-popular book has been reprinted numerous times and is now in the public domain. It contains Poincaré's informal but highly influential inductive definition of dimension.

Yaglom, I. M. *Geometric Transformations.* Translated by Allen Shields. New York: Random House, 1962. This book is aimed at a high school audience. Its goal is to consider elementary mathematics from the point of view of higher mathematics. Yaglom, who was a creative geometer as well as an enthusiastic teacher, wrote this excellent introduction to the concept of transformation. Highly recommended.

## INTERNET RESOURCES

Mathematical ideas are often subtle and expressed in an unfamiliar vocabulary. Without long periods of quiet reflection, mathematical concepts are sometimes difficult to appreciate. This is exactly the type of environment one does not usually find on the World Wide Web. To develop a real appreciation for mathematical thought, books are better. That said, the following sites are some good Internet resources.

Rusin, Dave. "The Mathematical Atlas 54: General Topology." Available online. URL: http://www.math.niu.edu/~rusin/known-math/indeex/54-XX.html. Accessed on March 28, 2010. A brief but well-written introduction to the most basic concepts of topology that also contains numerous links to other topics.

Eric Weisstein's World of Mathematics. Available online. URL: http://mathworld.wolfram.com. Accessed on March 28, 2010. This site has brief overviews of a great many topics in mathematics. The level of presentation varies substantially from topic to topic.

O'Connor, John J., and Edmund F. Roberston. "The MacTutor History of Mathematics archive." Available online. URL: http://www-history.mcs.st-and.ac.uk/index.html. Accessed on March 28, 2010. Probably the best site on the Web for biographies of mathematicians, the site also includes some interesting material on the history of various topics and an archive of "famous" curves.

"Point Set Topology." Available online. URL: http://coll.pair.com/csdc/car/carfre64.htm#OPENSETS. Accessed on March 28, 2010. This brief discussion of set-theoretic topology examines a number of important topological concepts as they apply to a five-element set, and for that reason, it can be very helpful to those starting out.

Watkins, Thayer. "Foundations of Point Set Topology." Available online. URL: http://www.sjsu.edu/faculty/watkins/topology2.htm. Accessed on June 7, 2010. This all-too-brief introduction is exceptionally well written. Too bad the author did not complete the effort. It is still worth a look.

## PERIODICALS: THROUGH THE MAIL AND ONLINE

### +*Plus*

Available online. URL: http://pass.maths.org.uk. Accessed on June 7, 2010.

A Web site with numerous interesting articles about many aspects of math. It has a free newsletter and a podcast. It is worthwhile to check occasionally to see what is on offer.

## *Scientific American*

415 Madison Avenue
New York, NY 10017

A serious and widely read monthly magazine, *Scientific American* regularly carries high-quality articles on mathematics and mathematically intensive branches of science. This is the one "popular" source of high-quality mathematical information that you can find at the grocery store or have delivered to your front door.

# INDEX

Page numbers in *italic* indicate illustrations;
page numbers followed by *c* indicate chronology entries.